AMERICAN STEEL

AMERICAN STEEL

RICHARD PRESTON

AVON BOOKS ◆ NEW YORK

Portions of this book, in somewhat different form, have appeared in *The New Yorker.*

AVON BOOKS
A division of
The Hearst Corporation
1350 Avenue of the Americas
New York, New York 10019

Copyright © 1991 by Richard Preston
Published by arrangement with Prentice Hall Press
Library of Congress Catalog Card Number: 90-20280
ISBN: 0-380-71822-7

The Prentice Hall Press edition contains the following Library of Congress Cataloging in Publication Data:

Preston, Richard, 1954–
    American steel / by Richard Preston.
        p.   cm.
1. Steel industry and trade—United States—Case studies.
I. Title.
HD9515.P76   1991
338.4'7669142'0973—dc20   90-20280
                                                                            CIP

First Avon Books Trade Printing: June 1992

AVON TRADEMARK REG. U.S. PAT. OFF. AND IN OTHER COUNTRIES, MARCA REGISTRADA, HECHO EN U.S.A.

Printed in the U.S.A.

OPM   10   9   8   7   6   5   4   3   2   1

For Doffy and Marguerite

Hast 'ou seen the rose in the steel dust?

—Ezra Pound

CONTENTS

■

PART I: IVERSON'S MACHINE

CONTENTS

PART II: **STARTUP**

■

PART I

■

IVERSON'S MACHINE

■

1

IVERSON

■ ▮ ■

September 25, 1987: Bulldozers hit the job site. There was no ground-breaking ceremony for the Crawfordsville Project, no time for that, this was an emergency. The Nucor Corporation, the ninth largest steel company in the United States, had purchased a square mile of cornfields south of the town of Crawfordsville, Indiana, a blank piece of Hoosier farmland upon which Nucor intended to restore American steel to its former greatness. At the southern border of Nucor's greenfield site, the bulldozers pushed up a wall of earth near the hamlet of Whitesville, to protect a farmhouse and a couple of bungalows from the thunder of melting steel.

F. Kenneth Iverson, the chairman, chief executive officer, and creator of the Nucor Corporation, proposed to install at the site a gigantic experimental machine for making sheet steel. No one knew if it would work, because a thing like Iverson's machine had never been built before.

Backhoes and excavators sliced into yellow Indiana earth, cutting a chain of holes twelve hundred feet long, for the roots of Iverson's machine. In a mountain range in West Germany, at a factory encompassed by tight security, master German craftsmen had begun to fabricate Iverson's machine. Iverson's machine consisted of one million parts. The machine was patented, and the idea for it came from a German inventor, Manfred Kolakowski, who was an employee of the German engineering firm that was building Iverson's machine.

The Nucor Corporation is a Fortune 500 company that makes fresh, plain carbon steel out of melted automobiles. The making of steel is said to be the most spectacular manufacturing process on earth. Steel is the monster of heavy industry; steelmaking machines are the fiercest, largest, and heaviest manufacturing engines on the planet; and there is money to be made in steel. For these reasons Ken Iverson believed that there is a place in the world for American steel. Iverson's new machine in Indiana was as long as a battleship. It was a starship for American steel. If it worked, it might grow Nucor explosively into one of the largest steel companies in the United States, if not into one of the premier steel companies in the world.

"We are going to leapfrog Japan," as Iverson put it.

■ ■ ■

On a warm day in March 1988, I paid a visit to the headquarters of the Nucor Corporation. The headquarters is located in Charlotte, North Carolina, in the Cotswold Building, beside the Cotswold Shopping Center, on the outskirts of the city. Soft Carolina clouds bubbled over the shopping center, and people wearing bright outfits strolled around the parking lot, getting in and out of cars made of steel. The Cotswold Shopping Center features a supermarket, some discount clothing stores, and a delicatessen called Phil's. The Cotswold Building, the site of Nucor's headquarters, sits across the street from the shopping center. The Cotswold Building is a four-story cube of buff bricks, glazed the color of dried vomit. The building is pierced with small aluminum-framed windows, and the entrance beckons with aluminum door handles that are shaped like bent paperclips.

I have always been attracted to industry, ever since I was eight years old, when I visited a Wonder Bread factory and saw loaves of Wonder Bread coming down the line. There was an American flag flying out front, and the people in the plant were covered with flour. That was in the sixties, when American industry was the greatest on earth. But when I pulled open the door of the Cotswold Building, I felt less like I was paying a visit to industry than I was paying a visit to the dentist.

I rode an elevator to the top floor, entered a hallway, opened a glass door on which was printed the word *nucor* in rounded letters, and found myself in the lobby of the Nucor Corporation. The lobby was the size of a walk-in closet. Nucor operates out of a suite of rented offices on half of the fourth floor. I gave my name to a receptionist and sat down in an orange motel-style chair. Nucor is said to have the smallest corporate headquarters of any Fortune 500 company. Nucor does business out of a rented office the size of a group dental practice. At that time, Nucor owned twenty-two manufacturing plants, flung all over the United States. The plants made steel and

■

steel products. There were only seventeen employees at Nucor's headquarters in the Cotswold Building, including secretaries and the chairman, Ken Iverson. So Nucor had a total of 0.8 corporate staffers per factory, including secretaries and the chairman.

The receptionist, Betsy Liberman, was piling tax manuals on a library cart. Since it was the month of March, income tax time, the Nucor Corporation would soon owe its usual millions to the federal government. In the hallway a schefflera plant in bad condition, a kind of starveling umbrella tree, seemed to block access to the Corporation. I heard a voice shout, "Hello! Ken Iverson! Nucor! Yeah, Iverson!" He was evidently talking on the telephone.

Promotional material on Nucor had been arranged in a rack in the lobby. I flipped through a sales catalog that described the wonders of Nucor bolts. Nucor's bolts are made from melted Cadillacs, among other lumps of recycled steel. I didn't care to buy a million bolts, so I turned to a Nucor annual report. The cover of the annual report displayed nothing but the names of all 4,500 employees of the Nucor Corporation, written in almost microscopic letters. The names were packed together, crammed into an endless sentence, without punctuation, without a verb, without a beginning or an end, a mysterious engraving that covered the annual report like a veil. A magnifying glass applied to a Nucor annual report would reveal the following names: KEITH E BUSSE WILLIAM W HAWLEY KENNETH D KINSEY JAMES E MCCASKILL MARK D MILLETT JANICE E ROACH DAVID C THOMPSON. These were some of the men and women who make steel.

The Nucor Corporation enjoyed a little less than $1 billion in annual revenues from the sale of steel and steel products, and earned $50 million in net profit, in a good year. The company's stock was trading on the New York Stock Exchange at around $40 a share. Nucor's main product was bar steel, slender lengths of plain carbon steel. Plain carbon steel is iron mixed with a little bit of carbon, which makes the metal hard, yet springy. Four of Nucor's twenty-two manufacturing plants were electric steel minimills, miniature steel mills that make bar steel from junk melted with electricity, and those plants both sold steel on the open market and provided steel to the rest of Nucor's manufacturing plants. Nucor produced 2 million tons of steel a year, made exclusively from trash, and Nucor's sales force sold the metal for around 17 cents a pound, with a net profit of a penny a pound or $20 a ton—a good margin, for steel.

Steel isn't such a bad business, anyway. Americans consume the equivalent of five Great Pyramids of solid steel every year. That's 105 million tons of steel, if you include the steel in imported automobiles, a cool $40 billion worth of metal. Close to half of it is recycled steel. Toyotas, tin cans, skyscrapers, bolts, and supertankers, they wind up back in a furnace in a steel mill.

Iverson's phone conversation continued for a minute—"Yeah, that's fine. O.K.!"—and it ended with the clatter of a telephone being hung up. There came a sound of footsteps in the hallway, and F. Kenneth Iverson appeared.

The chairman and CEO of the Nucor Corporation was in his sixty-third year, and he had a tall, limber, rangy frame, and pale gray luminous eyes. He wore a starched white shirt, no jacket, a dark silk paisley tie, and wrinkled cordovan shoes. A modest gut bulged his shirt. He led me down a hall.

Iverson settled himself behind his desk, in a corner office that faced west over the shopping center but had no view because venetian blinds were drawn. His office was the cheapest-looking executive lair I had ever seen. His veneered desk was worth all of $300 and it had a black vinyl top. The walls of the office were coated with tan vinyl meant to resemble leather. There were a couple of bookshelves, and one of them held a massive tome, *The Making, Shaping, and Treating of Steel*—the steelman's bible. I sat down in a chair. It was covered with textured plastic meant to resemble suede, but it didn't feel like suede, it felt like a scrub pad.

"There's nobody at this company who's covering their ass with paper," said Iverson. He knocked a Winston Light out of a cigarette pack and put it on his lip. He picked up a yellow plastic lighter from his desk, hit the cigarette with flame, and inhaled a chestful of Light smoke, holding the cigarette gently in his fingers. Iverson's gray eyes looked at me through smoke.

"We have people come in here who think we're the real big shots in the company. It gets me irritated. We don't do much here at Charlotte," he said. "That's not a joke. Except the cash! We handle the cash!" He grinned, and a pressure wave exploded across his face, wrinkling a set of crow's feet at the outer corners of his eyes, exposing a chipped upper front tooth, and pushing his ears away from his head. He had large elliptical ears and blond graying hair, parted at the side, short on the temples and brushed up on the forehead. You could imagine that his eyes were the silvery gray of plain carbon steel.

"The Big Steel companies in the United States are scared," he said, in an affable voice. "They don't *say* so, but they are scared." He tapped his

■

Winston Light into a soapstone ashtray. "We're going to get into the market for flat-rolled sheet steel with our mill at Crawfordsville. That's the largest single market for steel in this country." Iverson had the kind of breezy, sociable voice you hear in businesspeople who have spent their lives doing fruitful business with every twisted life-form that comes down the pike in a human shape.

There was a photograph of a black swan on the wall. "That's Jessica. I raise waterfowl. Swans are mean, most of 'em. Black swans have a better disposition than white swans. Jessica's my favorite." On a bookshelf there was a steel hubcap marked with the word *Reo*—"That came from a Reo truck," he explained.

Nucor is the successor to the Reo Motor Car Company, founded in 1904 by Ransom Eli Olds, the inventor of the Oldsmobile. Reo Motor Car had become a company called the Nuclear Corporation of America, and that had mutated into Nucor, and Nucor had proceeded to melt Oldsmobiles in such numbers that they might have been snowflakes hitting warm pavement. The Reo hubcap on Iverson's shelf was the only hard asset of Reo left on Nucor's books.

Nucor, née Reo Motor Car Company, had burst into the steel business unannounced and uninvited, and F. Kenneth Iverson, the architect of Nucor's arrival in steel, had become something of a cult figure in the international steel industry. Iverson was in certain respects the most carefully watched steelmaker in the world. Steelmakers from Japan, Korea, India, Brazil, and Turkey asked each other at steel conferences, in business English, "What is Mister Iverson *doink* now?" Executives in the American steel industry—at the so-called Big Steel companies, like the USX Corporation and the Bethlehem Steel Corporation—often asked themselves a similar question—what in hell is Iverson doing now?—and it drove them crazy just to think about it. Iverson had ripped a billion dollars' worth of steel markets away from Big Steel. All of Nucor's plants were nonunion plants. Iverson was not a popular man with the United Steelworkers union. Iverson had built his steel mills and steel product factories in small rural towns, among farm people who were comfortable with machinery and hated labor unions. On their days off, some Nucor steelworkers continued to farm a little corn and slop a few hogs. On occasion they offered to recycle a union organizer into a bar of steel.

Iverson's net worth ran somewhere between $15 million and $20 million, depending on the price of Nucor stock. He was a mechanical engineer and a metallurgist by training. He described himself as a hot metal man. In

∎

the mind of a hot metal man, steel in its natural state is a river, continually changing its shape and yet always the same thing, and the river shines like the sun.

"Hot metal has a fascination all its own," as Iverson explained it. "There is a fascination about melting a metal and pouring it into a shape. A hot metal man has to feel that fascination. That's why he works around hot metal."

Iverson tried to visit each of Nucor's plants and steel mills at least once a year, where he walked the full length of the plant's manufacturing line with the plant's general manager. Therefore Iverson walked the total combined length of all of Nucor's manufacturing lines once a year. The chairman and CEO of the Nucor Corporation walked the line with a hard hat on his head, with a pair of dark steelworker's glasses clipped to the brim of his hard hat, and he wore a green spark-proof jacket, a furnace man's jacket. At times he wore a hearing aid in his left ear, because he had gone a little deaf from hanging around too many monstrous machines.

Iverson enjoyed the carnival in a detached kind of way, with his hands in his pockets, bathed in the hoots of sirens and the concussions of the outstreaming tons as they moved through rolling machines, *ba-boom, ba-boom, ba-boom*, the sound of glowing strips of bar steel nosing through a rolling machine at thirty miles an hour, *ba-boom, ba-boom, ba-boom*, filling his longbones with the heartbeat of American steel.

The metal running through those machines wasn't hot: It was only red-hot steel, at a temperature of 1,900 degrees Fahrenheit, as hot as a bed of coals, as hot as hell, and that wasn't hot to a hot metal man. Steel in its natural state doesn't qualify as hot until it's nearly 1,000 degrees hotter than that, until it's close to 3,000 degrees Fahrenheit, until it's painful to look at with the naked eye, until it's as runny as water and as unpredictable as a cat. Following his inclination for hot metal, Iverson wound up in the steel mill's melt shop. A melt shop is a building as large as a cathedral. In a melt shop, shredded automobiles, sheared industrial machinery, and blocks of steel trash are melted inside electric arc furnaces, huge kettles for cooking steel.

An earsplitting roar was coming from a furnace. A crew of hot metal men, drenched with sweat, hurried back and forth in front of an open door in the side of the furnace. A beam of light, like a spotlight, emerged from the door. Inside the furnace there was a bubbling bath of live liquid steel, known to hot metal men as a heat of steel. Certainly it was live steel, the metal was churning with dissolved oxygen. Iverson would notice a steel-

■

worker he knew and he'd yell something like "God damn! James McCaskill! I didn't know you were working here! You'll go anywhere for this company!"

"Is that you, Mr. Iverson? Well, God damn! I didn't recognize you!"

Feeling the fascination, Iverson stood close to the furnace with his hands in his pockets, contemplating the changes of hot metal. A restless crawling fierce white light, tinted with orange, streamed out through the furnace's door and fell over Iverson and the hot metal men, throwing human shadows on the wall of the steel mill. The light of hot metal is bright enough to varnish your face with a first-degree burn, like a sunburn, except it is a steelburn. Iverson reached up to the brim of his hard hat and flipped down his pair of dark steelworker's glasses. Through the smoked glass he could see the structure of the heat, shining inside the furnace.

"The steel almost looks—if it wasn't for the heat—it almost looks innocent," in Iverson's words. "You have to be aware of how wild it can be."

Seen through dark glasses, a heat of steel had a weird structure. If the heat was calm, the molten metal displayed a shiny, reflective surface that was as smooth and wet as a puddle of mercury. At the same time it glowed with an inner light. You wanted to put your hand into it, just to see what it felt like. You wanted to scoop out a cupped hand full of steel and let it dribble through your fingers. If you were to thrust your hand into a pool of steel your hand would explode. Your hand would blow up in your face and spatter you with carbon steel, setting your clothes and hair on fire. To Iverson, hot metal looked like a pool of cold water. The steel rippled with a dull, silvery glitter inside the furnace, obviously the color of plain carbon steel, and yet it glowed. It threw off a blackbody light, a light from iron atoms bumping around. The metal was at a temperature of 2,900 degrees Fahrenheit, the bones of civilization melted into a pond, as reflective as water and as incandescent as the dawn, brooding with probabilities. The metal swirled gently, as if it were being pecked by trout, and curdles of snow-white limestone slag floated on the metal, like blobs of ice drifting on a pond when the ice is going out at the end of March.

Iverson could smell the heat. It had a musky biting odor, like the tang of a dead skunk on a summer night. He inhaled the unmistakable odor of a heat of steel, a mixture of vaporized iron, carbon smoke, slag, and cooking firebrick, a profound blend of odors once smelled never forgotten, an addictive gas that gets into the blood of a hot metal man.

The metal looked so sweet. But you couldn't forget that you were

■

looking at live steel. Live steel can come to a boil in a flash. The term for that is a carbon boil. A severe carbon boil can shake a steel mill to its foundations. Something horrible churns against the furnace's walls, trying to escape, throwing steaming hot metal out the door. Ambushed by a carbon boil, Iverson backed away from the furnace, sparks monsooning around him, drowning his outline in fire, and then he was just another hot metal man lost in a hot metal rain, with eyes like two black coins, staring into the blackbody light.

■ ■ ■

Iverson reached up to a bookshelf in his office, pulled down a blueprint, and laid it on his desk. The blueprint showed the outlines of the Crawfordsville Project, in Indiana. He moved his finger over the plan, tracing the shapes of three gigantic buildings, each the size of a skyscraper lying on its side.

"The Crawfordsville Project is immensely complicated but immensely simple," he said. "We are constructing one million square feet of building space. This is a unique process. We're going to cast a very thin slab of steel, two inches thick and fifty-two inches wide, then we're going to roll it down into a sheet, in one continuous process."

Iverson's machine, the heart of the steel mill, was the world's first continuous steelmaking machine. It was known as the Compact Strip Production facility, the CSP. Iverson's machine was 1,177 feet long. Two buildings would house the machine, because it was too long to fit inside one building, even a mill building as long as a skyscraper lying on its side. The product of the CSP machine would be hot-rolled sheet steel, also known as hot band steel, so called because the metal is rolled flat, into a band, while it glows the color of a tangerine: hot band steel. It is a heavy sheet steel, used in truck frames, machinery, and water tanks.

The third building was a cold mill. The cold mill would contain machines that would cold-roll some of the metal that came out of the CSP, elongating the hot band steel, making it nearly as thin as paper and putting a shine on it: cold-rolled steel. It is a thin, smooth sheet steel, used for automobile fenders, filing cabinets, and small parts inside computers.

"The Rust Belt is the reason we placed the machine in Indiana," Iverson said. "Indiana is in the center of the Rust Belt, and the Rust Belt has a huge excess of scrap." Iverson's machine was going to make one million tons of steel a year from busted junk. Twenty-five thousand railroad cars a year would deliver the Rust Belt to the Crawfordsville Project. The

■

railroad cars would carry shredded automobiles, factory punchings, cut-up steel mills, that sort of thing. The Rust Belt would be cooked in twin electric furnaces.

Then the resulting molten steel would be drained out of a furnace, carried in a ladle up to the head of the Compact Strip Production machine, and poured into a copper funnel at the head of the machine, and a thousand feet down the line, along would come sheets of rich, glistening steel. The steel would be coiled up with a snap, like a roll of paper towels, and from there it would be moved up a conveyor ramp and out into the Midwest, to be sold by Nucor's sales force. Iverson's machine would turn the wreckage of America's industrial decline into a ribbon of brand-new steel. This was the resurrection of the Rust Belt.

The machine would run on three levels of software; and as the software was perfected during the operation of the machine, the machine would transform itself over time into something of a robot.

"In the steel industry, you get a development like this once every fifty years," he said.

Although Iverson's machine was as long as a battleship, it was a gadget in comparison to a major steel mill. A major steel mill sprawls from two to seven miles along a body of water. The making of steel, in the traditional method, takes place in steps, in a manufacturing process that was developed in the nineteenth century and hasn't changed very much since then: coke and iron ore go into a blast furnace and out comes liquid pig iron. The liquid pig iron goes into a steelmaking furnace and out comes liquid steel. The liquid steel is poured into pots to make solid ingots of steel, or, in a more recent innovation, it is poured into a continuous-casting machine, which extrudes a solid slab of steel. The ingots or slabs are allowed to cool to room temperature. Then they are reheated, red-hot, and driven through a rolling mill a half a mile long, and out comes sheet steel. The method requires machines that cost billions of dollars, it burns huge amounts of energy, and it depends on the mass labor of thousands of steelworkers. That is to say, the manufacture of steel is capital intensive, energy intensive, and labor intensive. The Nucor Corporation intended to do for the steel mill what the Apple Computer Corporation had done for the computer. Iverson intended to cram a steel mill into a box. Iverson's machine was a desktop steel mill.

There had been a terrible accident in the American steel industry. American steel had been knocked flat and almost died. Hundreds of thousands of steelworkers had lost their jobs, and many steel mills had been closed

■

forever. When Americans lose, they lose big, and American steel had lost spectacularly in world markets. In simple terms you could say that Americans had failed in the steel business. Or rather, they had almost failed. The American steel industry had not yet given up hope.

Iverson's machine was supposed to manufacture one million tons of steel a year with only five hundred steelworkers. In Japan it requires 2,500 steelworkers to make a million tons of steel a year. So Iverson figured that the machine was going to be five times more labor-efficient than the Japanese steel industry, which is regarded as being one of the most advanced steel industries in the world. In Japan it takes four or five man-hours of labor to make a ton of steel. "We're hoping to get to *one* man-hour per ton of steel at Crawfordsville, and that's including everybody, even the guy who sweeps the floor," he said. "I don't *care* about the cost of labor in Korea or Brazil. The Brazilians could pay their people a dollar an hour, it wouldn't make any difference to us."

I asked him if he was scared.

"Scared? What—scared it won't work? I'm comfortable with it." He puffed his cigarette. "I'm *enthusiastic* about it."

He had played the baritone saxophone as a teenager, and his voice seemed to have absorbed a metallic energy from blowing deep, grunting swing.

"In Crawfordsville, where almost everything is new, I have young guys," he went on. "Young people can accept challenges. Young people can accept new ideas, whereas older people are trying to protect their hard-won gains. We expect our people in Indiana to make mistakes, because whoever you put in a system like that is going to make mistakes." Headhunters were busy worming through the guts of the largest American steel companies, hiring young manufacturing wizards away from Big Steel for the Crawfordsville Project.

Iverson wanted to build the steel mill in eighteen months, breakneck speed. He would crash the mill through fast-track construction, in which a plant is designed while it is being built. He had originally budgeted $225 million for the Crawfordsville Project, but not surprisingly, the budget had grown larger, and recently it had hit a quarter of a billion dollars. Much of it was borrowed money: commercial paper, short-term debt. He had handed a quarter of a billion dollars to a bunch of young people and had told them to go build a steel mill that would blow the hair off the world steel industry.

"I've got an interesting report, right here, by Bethlehem Steel," he said. "They're watching us. It's Bethlehem's opinion of our program at Crawfordsville. It's Bethlehem's evaluation of why it won't work."

■

He stood up from his desk and fished a manila folder from a bookshelf. "The report is *an inch thick,*" he said in a tone of wonder. The folder hit his desk with a smack.

"Has it been published?" I asked.

"No!" His eyes flickered with amusement as he sat down. "This is an internal memorandum. Somebody gave it to us."

Iverson had friendly arrangements with traitors inside Big Steel.

"That Bethlehem Steel would expend this amount of dollars and effort to draw up a monstrous report to say that our Compact Strip plant *won't work* is just amazing," he said. "They've got cost models in here! They say here that we couldn't possibly build one of these plants for less than four hundred million dollars! And even if we did build it, they say we wouldn't have any competitive advantage!" His grin widened until it was positively huge, exposing his chipped tooth, pushing his ears away from his head.

According to the folklore at the Nucor Corporation, about the only thing that Walter F. Williams, the chairman of Bethlehem Steel, would allow his engineers to build these days was a stack of paper describing why a machine won't work. According to the Nucor theory of history, none of the large American steel companies had the imagination or the will to commit capital to build a new machine.

"Almost all the developments in the last thirty years in steel have come from outside the United States. We became complacent," Iverson said. "Oh, yeah, there's resentment of me. I hear about it. I get it played back to me. Think of it from a Big Steel executive's point of view. If I had spent my life in steel, if I had worked my way up to be a vice-president at Bethlehem Steel, and these little pip-squeaks like Nucor came along and ruined my business, I'd be resentful, too. It is now apparent to them that the American minimills are at least as much of a threat to them as imported steel. It's taken them about twenty years to realize it. Foreign steel came in the back door and we came in the front door of their snug little house."

2

THE MARSHAL AND
A FEW OF HIS BOYS

■ ■ ■

A clear dusk in May 1988, in Indiana. The sun has set over the town of Crawfordsville, leaving a rusty glow on Victorian facades of brick buildings along Market Street. Some of the stores are vacant, their windows whitewashed on the inside of the glass. Crawfordsville, population 13,000, is the county seat of Montgomery County. Crawfordsville sits in the backyard of Big Steel. A hundred and twenty miles north of Crawfordsville extends the southern shore of Lake Michigan, the Lake Shore, where the best steel mills in America are situated.

The first white settlers around Crawfordsville came down the Ohio River on flatboats loaded with hogs, oxen, furniture, and people. When they came to the mouth of the Wabash River, at the southern tip of Indiana, they stopped and cut long poles of mockernut hickory. Then they poled their flatboats upstream along the Wabash, a slow and torturous job. Two hundred miles up the Wabash, some of them turned northeast on Sugar Creek. Sugar Creek in those days was ten feet deeper than it is today, and perfectly transparent. They poled up Sugar Creek, their hickory wands bending through deep clear water onto boulders of jewel-like colors—red, blue, green—that passed slowly under their flatboats. From boulder to boulder shot huge shadows of spooked bass. Sycamore trees as big as worlds extended over the water. The sycamores had boles that were twenty-five feet in circumference, and their crowns flashed with ruby-throated hummingbirds. In sandbars the pioneers found small diamonds and flakes of placer gold. Tallgrass prairies opened on the north side of Sugar Creek,

■

inhabited by bison. The black prairie soil was so rich you could burn it like coal. To the south of the creek stretched a country of yellow soil, covered with a climax forest dominated by sugar maples and beeches of prodigious size. In the early days they called that country the Big Woods; it is mostly farmland today. Crawfordsville was settled in 1824 on a bluff overlooking Sugar Creek.

A softball game breaks up on a diamond down by Sugar Creek. At the McDonald's, beside the shopping mall to the south of the town, close to where the Big Woods once began, weary parents try to calm their shouting children. A Pizza Hut across the street collects brash alert teenage girls with moussed hair that looks like it has been rubbed with fat and electrocuted. There are several manufacturing plants in Crawfordsville—the Norcote Chemical plant, the world's only plant that makes the ink that is bonded to the surfaces of compact disks; an R. R. Donnelley & Sons printing plant, the third-largest book-manufacturing plant in the world; and a factory called California Pellet, which makes the machines that make dogfood pellets. The factories operate deep into the night, making useful objects with machines.

Crawfordsville calls itself the Athens of Indiana. General Lew Wallace, a commander at the battle of Shiloh and author of the novel *Ben Hur,* lived in a mansion near the center of town. Crawfordsville is served by the Monon Line, a railroad linking Louisville to Chicago. The Monon carried Abraham Lincoln's body slowly through Crawfordsville. They say that General Lew wrote parts of *Ben Hur* while riding the Monon Line. Perhaps he imagined the chariot race while riding the Monon and watching little Hoosier towns go by.

One block off Main Street, near the center of town, a plastic sign that looks like a glowing pillow hangs over the door of the Scoreboard Lounge, inviting seekers of liquor and a game of pool to enter and make themselves at home. Keith Earl Busse, the vice-president–general manager of the Crawfordsville Project, sits inside the saloon at a Formica table, with a group of Nucor managers and steelworkers. The air is thick with rock-and-roll music and shouts. It is a bachelor party for a steelworker.

■ ■ ■

"Let me have a CC & 7," Keith Busse said to a waitress.

"A *what?*" she shouted over the music.

"A Canadian Club and Seven-Up. Tall, on the rocks."

■

15

Busse leaned forward with his elbows on the table and looked at me with a pair of green eyes. "I think Rome's on fire and we don't see it. Sometimes it seems that everybody coming out of school wants to go to Wall Street or be a lawyer. Everybody in this country is a *service* individual. We're a nation of ambulance chasers! How can you fix something if you don't know it's broke?"

His CC & 7 arrived, and he tasted it.

From a poolroom beside the bar came a clatter of pool balls and a yodeling scream. Some of the Nucor steelworkers referred to themselves as the Run-a-Muckers, and that was what they were doing at the Scoreboard Lounge.

"No one in this bar has ever built a steel plant before," Busse went on. "Most of 'em have never done construction of any kind. What better way is there to learn? The mind's a powerful thing."

A voice shouted from the poolroom: "Let's puke Chris! If Chris doesn't puke tonight he should be fired!"

"Nucor is not afraid of youth," says Busse. More screams out of the poolroom. Over the racket, Busse explains that the Run-a-Muckers will manage the construction of the steel mill. They will hire and supervise construction firms to build the steel mill, and then, after the mill is built, the Run-a-Muckers will tune up the machines and make steel. "If Big Steel really knew what we are capable of. If they only knew! They know, but they don't know. They pretend like we don't exist. They pretend like they aren't worried about us."

Keith Earl Busse was something of a mysterious figure in the American steel industry. Few people in the industry knew anything about the man. It is possible that he was a marshal wearing a steel star, sitting in a saloon drinking whisky, and pretty soon he and his boys were going to clean up the American steel industry. Keith Busse was a stocky person, five feet ten and a half inches inches tall, born and raised in Fort Wayne, Indiana. He termed himself a Hoosier. No one really knows where the word *Hoosier* came from. Busse's last name rhymed with "fussy," like this: *Buh*-see, with a short *u* and a long *e*. He was the offspring of German Midwesterners, believers in the God of Martin Luther and in the value of Work with a capital *W*. Busse had a square face with a firm jawline. His hair—brown, bushy, stiff, and neat—fell in fluffs over his forehead. Busse's green eyes looked straight through people, as if his eyes were emitters of a high-energy radiation. He wore a pressed button-down shirt of a pale-green hue that matched the color of his eyes. He wore Lee jeans and steel-toed boots.

A mild belly deployed over Busse's jeans. His arms were sinewy and thick. He was forty-five years old. "I'm the old man around here," he remarked, glancing down the table at his employees, who were in their twenties and early thirties; most of them were male.

Keith Earl Busse was a manufacturing wizard. He knew something about the design, construction, and operation of factories of the future. He had designed and built the only operating standard-bolt factory in the United States, a Nucor plant. The Nucor bolt plant, in Saint Joe, Indiana, is staffed with ex-farmers and robots. The farmers run the robots. Virtually all standard bolts in America other than Keith Busse's bolts are imported. A Busse bolt has a small letter *n* stamped on the head, which stands for "Nucor." You can find Busse's bolts at your local hardware store. Find the little *n*, and you know you're buying a bolt made from natural clean native busted automobiles.

Another of Keith Busse's specialties was negotiation. Busse would negotiate anything, whether it needed to be negotiated or not. He had a known tendency to start arguments with Ken Iverson that could turn into caterwauls at the Cotswold Building, until Busse had got himself worked up enough to pace a room, slamming his fist into his palm, saying of the chairman, "God damn Ken Iverson! He pisses me off sometimes!"

Keith Busse was an accountant by training, a sharp-pencil boy with a background in bolts, not a hot metal man. To a hot metal man, your bolts are not your hot metal, your bolts are little cylinders chopped from steel rod, bearing no resemblance to live liquid steel. In the steel industry you are either born with steel bonded to the hemoglobin in your blood or you are considered to be a nobody. Keith Busse was almost a nobody.

He also happened to be the biggest machine gun dealer in northern Indiana. He ran a gun supermarket in Fort Wayne. That made a mildly favorable impression on the hot metal men, those few who happened to know that Busse was selling machine guns. It wasn't enough to convince them that Keith Busse knew anything about steel, but machine guns were a step in the right direction. The gun trade was only a profitable sideline for Keith Busse that had nothing to do with the Nucor Corporation. If Keith Busse wanted to sell machine guns in his spare time, that was all right by Ken Iverson.

Busse sipped his whisky and pop and insisted that he wasn't personally worried about anything. "I've got a gun store up in Fort Wayne," he declared. "I own it with a partner. My partner and I built it. Doing this mill is just like building a gun store, except it's bigger in scale. With this

■

steelmaking technology we've got, it's gonna be Big Steel fighting a war against us with a bolt action rifle when we've got a machine gun. But I don't know. Perhaps we are just a pimple on the camel's butt. I don't want to be cocky. We'll do the mill first and talk about it later.''

Busse rattled the ice in his glass and ordered a second CC & 7. It arrived quickly, and he tasted it.

■ ■ ■

A young guy sat down at the table near Busse with a mug of beer in his hand. ''Wot's happening, Keith?'' said the young guy, with a British accent.

''I'm about to clean some people out,'' replied Busse. Busse stood up and said to me, ''He's an Englishman. We found him as a stray dog. He was wandering around this country with a knapsack.''

Another scream flew out of the poolroom.

Busse said to the Englishman, ''I'll see you later. I'm going to win some money.'' Busse disappeared into the poolroom, into a wave of anarchic shouts: ''Yo, Mr. Busse! Yo!''

The Englishman was Mark D. Millett. He was the manager of the melt shop, one of the three large buildings at the Crawfordsville Project. He was twenty-eight years old, a native of Plymouth, England. Mark Millett was a wiry person, below average height, with a fatless body, a suntan, and brown, floppy, longish hair. Millett's eyes were dark brown and large, almost too large; restless, intelligent, rather guarded eyes. Veins and tendons stood out crisp and sharp on Millett's hand, which was wrapped around the handle of his beer mug.

He had earned a bachelor's degree in metallurgy at the University of Surrey, outside London. After graduation, he went to the United States for a year, ''to be a ski bum and to party,'' as he put it. He encountered Abby Trowbridge, of Brooklyn, New York, and they ended up living in Aspen, Colorado, in modest circumstances and in love. Meanwhile, Mark drifted from one dishwashing job to another until he found a job as a part-time cook at a ski lodge.

''I'd stuff my face on the job and then starve for two days while I skied,'' he said.

He and Abby were married, and he got a job with Nucor as a metallurgist. He cooked his face with first-degree burns by staring at liquid steel. ''You come to earth in a severe bump,'' he said, referring to his first good look at liquid steel.

■

The melt shop, of which Mark Millett was the boss, would handle all phases of liquid steel. The building would contain the twin electric furnaces (for melting steel), a metallurgy station (for refining steel), and the upper portion of the Compact Strip Production machine, known as the casting tower.

The big mystery inside the CSP was a machine inside the casting tower, a module at the head of the CSP. It was a thin-slab casting machine with a copper funnel. This casting machine was an experimental unit. If it worked, then an endless thin slab of steel would emerge from the casting tower, like a strand of fettuccine two inches thick, and from there the slab would be crushed and elongated into a sheet of steel.

Inventors had been trying to invent a machine that would make an endless strip of steel since 1856, when Sir Henry Bessemer had tried it and failed. There had been numerous inventions since then, all kinds of contraptions. Uncounted millions of dollars, pounds, francs, rubles, and yen had been poured into cuckoo clocks for making a ribbon of steel, to no avail. All the machines had come to universal grief in jam-ups, meltdowns, conflagrations. There had been so many failures. The hot metal men had underestimated the cruelty of steel. To make a thin sheet of steel with one machine was the single biggest unsolved problem in the manufacture of steel, and it had been the biggest problem since the days of Queen Victoria, in the middle of the last century. Any company that could solve the problem would by definition become the global leader in the manufacture of steel. In the long glare of a century and a third of burning machines, F. Kenneth Iverson and Keith Earl Busse had given Mark Millett, an English ex-cook, twenty-eight years old, the assignment to put metal into the machine and to make it work. If something went wrong, the casting machine might literally blow up in Millett's face.

"Probably the caster will be the hardest part of the plant to get up and running quickly," said Millett, in a somewhat casual way.

A gaunt, bearded giant of a man slammed through the door of the Scoreboard Lounge. "*Ooooooo,*" he said in a low voice, heading for the poolroom. His passage through the saloon provoked dead silence.

Millett's head swiveled. "Did you see that guy?" he said. "He's a steel hanger. Those are the guys who hang the girders and put the bolts in them. He's been putting together the melt shop with his crane."

Millett poured himself a fresh beer from a pitcher and glanced into the poolroom. "Keith Busse is likely to win that game," he said. "If Busse gets behind, he bets fifty dollars. He keeps doubling until he wins, and then

■

he shuts down the game.'' Millett grinned. ''I'm going in there to see if I can't win some Busse bucks.'' Millett took up his beer and wandered into the poolroom.

■ ■ ■

In the poolroom, Keith Busse was pacing up and down with a pool cue in his hand, working on another CC & 7. Busse had placed yet another CC & 7 in reserve on a cocktail table, a long, tall, final slam of Canadian whisky and soda pop.

Sitting on plastic couches and leaning against the wall, Nucor steelworkers were watching the pool game. Busse had already won money off his employees, who seemed to be amused by their own misfortune.

Busse said, ''Let's have a few games of pride, no money,'' while he racked and broke the balls.

The bearded steel hanger stood at the center of a knot of steelworkers, swaying above them like an erection crane. He had a long nose. He was a foreman for a contracting firm; he was a crane operator. He pulled a chrome flask from his pocket and upended it to his lips. The flask glittered in the ceiling lights. He put the flask back in his pocket. He held the liquid in his mouth, savoring it, and then swallowed it, and his nose twitched, as if he were trying to clear it of an obstruction. He dragged on a cigarette and began to move through the crowd. Suddenly he stuck a prehensile finger into a steelworker's chest, and said, ''My name is Duane Hurler. I hang iron. I am fucked up on moonshine, and I am going to kill you.''

The steelworker backed away. ''You'd better talk to my supervisor.''

''Your supervisor! He ain't worth a fuck anyway. I am going to kill him soon as I kill you.''

''I'm sorry—''

Duane Hurler's nose twitched again, as if he were still trying to clear it of an obstruction, and he backed the ''steelworker'' toward the wall. (Hurler's name has been changed here.) ''You Nucor fucks don't give me anything I ask for!'' Hurler said. ''When I say I want some concrete, you don't give me any *concrete*. I have not been able to hang iron for a week! Because you little Nucor fucks won't give me any concrete. I am going to go home and raise coon dogs if I cain't hang some iron soon. I am going home to my honey in Arkansas. I really am. I really am. I am going home to my honey and my coon dogs. And first I am going to kill you, you little sucker and you think that's funny and you are a little sucker.''

The steelworker was a stout fellow in his twenties, with gentle eyes

■

20

behind eyeglasses, who had never been inside a steel mill in all his life. His name was Kenneth D. Kinsey. He kept saying, "No, really, I'm sorry, Duane," as the iron hanger backed him around the room.

No one paid much attention to the iron hanger and the frightened steelworker. Keith Busse kept asking people to move out of the way so that he could make a shot with his pool cue. "Excuse me. Excuse me. Three in the corner pocket."

A bachelor named Chris, the excuse for the party, sat at a cocktail table drinking something called a Nucor, which was a water glass filled to the brim with a mixture of scotch, gin, tank-car vodka, rum, tequila, and a splash of Jack Daniels, stirred gently, and cooled with one cube of ice. The drink had the word *projectile* written all over it. The bachelor was as marble-eyed as a pithed frog, without any life left in him but a beating heart.

A short, quiet, soft-spoken "steelworker," with a sunburned forehead and a red beard, remarked to me, "Us Nucor guys have to build the damn steel mill before we can use it. They hired me off a farm. I was a farmhand. My dad's a farmhand. I've lived in Montgomery County all my life. There could be some pretty good money in this steel mill, once they get it runnin' right."

Busse finished his games of pride and began to lose a pool game to Mark Millett, the melt shop manager, and there was money riding on the game.

"I'm doing all right tonight," said Millett, turning to the crowd. "I'm up five bucks. Busse bucks."

"Yeah?" said Busse. "I'm going to have me some Millett money pretty soon. I took Mark Millett for fifty last time. I'm going to do it to him again, only worse. Two in the corner pocket." Busse drove a shot at the two ball, but it missed the pocket. Busse backed away from the table, looking disgusted.

"Busse bucks," said Millett, glancing at Busse. Millett stroked chalk on his cue and sank a ball. Millett missed the next shot.

Busse leaned over the table. "Millett money," he said. "Nine in the side pocket." He stretched low and sank the nine on a rebound off the cushion. "Ten in the corner pocket." Busse sank the ball in a side pocket.

"Keith, what pocket did you say?" said Millett.

Busse didn't reply. He grinned and grabbed for his long, tall CC & 7, the deep reserve on the cocktail table.

Millett sank a ball on a double bounce, and then sank two more balls, clearing the table.

■

Busse smiled and pulled out his wallet. He waved a twenty-dollar bill before the crowd. "That's the first time I've given money to Mark."

Millett shook the money in the air. "Busse bucks!" he said, and the steelworkers grinned.

Hurler offered his chrome flask to Busse, to console him. "You try a little of that," he said.

Busse accepted the iron hanger's flask, sniffed at it, took a swig, and gasped. "That's absolute poison," he said, wiping his mouth on his bare arm.

"Yes. That really is good." Hurler inhaled from his cigarette and his nose twitched again. "That was distilled in Mountain Home, Arkansas," he said. He swung around and offered the flask to Ken Kinsey, the intimidated steelworker. "Try a little of that. No, it's good. It really is good. You try that. It will help the fluttering."

Kinsey put the flask to his lips and exclaimed, "Ah!"

"Yes, you *tell me* that's good." A white grin rippled through Hurler's beard. "I like moonshine." His voice got louder. "I hang iron." He gave testimony to a clump of steelworkers. "I started hangin' iron when I was seventeen years old. I have been twenty-six years in the biniss. My daddy got me into the iron-hangin' biniss. I tell you one thing about iron-hangin', you learn pretty quick how to get drunk and fuck up. But I don't ever look in a rigger's book and I have *never* turned over a crane! I work for Red River Steel!" (The name has been changed.) Long as I work for Red River, I *am* Red River Steel! I can beat anybody at one thing, and that's hangin' iron! *Ooooooo!* There ain't none of these steel erection firms around here that can *touch* us for steel erection! The only thing will give you a proper erection is the way we do it at Red River Steel!"

3

CRANE DROP

■ ■ ■

The day after the night at the Scoreboard Lounge, I drove south from the town of Crawfordsville to see the Nucor job. The Crawfordsville Project was located four miles south of the town on County Road 400 East, a ribbon of potholes that headed straight into fields and woods. Soft grass blew on the roadbanks, and stands of tall timber put out leaves in clouds on the land. County Road 400 dipped through a pocket of woods where maples were blooming with green flowers, and crossed a creek on a steel bridge. Sycamores grew along the creek. The road came out of the woods and climbed a rise, passed a development of contemporary ranch-style homes ("Chigger Hollow, by Leland Stark"), and entered plowed farmland.

The road ended at a barbed-wire fence and a gate where there was a guard trailer. A security guard emerged from the trailer, a young woman, and she asked me my business. She handed me a small piece of paper signed by Keith E. Busse, a site pass.

I drove straight ahead from the guard trailer for almost a mile over a tormented, half-dried mud surface. My car approached three clusters of steel columns, the skeletons of the three enormous mill buildings that made up the Crawfordsville Project. They had burst from the ground during the winter, a crop of winter steel. Each building had its own manager, as I understood, who reported to Keith Busse. Mark Millett ran the melt shop, a certain Rodney Mott ran the hot mill, and one Vincent Schiavoni ran the

■

cold mill. Each manager was racing the others to see who could complete his building first.

The steel columns sat on concrete piers. Erection cranes towered over the buildings. At the bases of the buildings, clouds of dust were boiling, and a throb of diesel motors vibrated the air. I threaded my car around dump trucks, bulldozers, and road graders, all in crisscross motion around my car. The earthmoving machines lumbered here and there among the steel structures, trailing williwaws of dust.

The only finished result of the crash, so far, was a small one-story building made of cinderblocks sitting off to one side. This was the Admin Building, as they called it, where Keith Busse had a corner office. Busse's office breathed a sharp smell of plastic wallpaper. On a coffee table sat a chromed Nucor bolt as thick as a truck axle. The bolt was made of solid steel and weighed a good thirty pounds.

"We've dreamed of this mill for a long time," said Busse, leaning back at his desk and throwing a mud-caked boot over one knee. He inserted a coffee stirrer between his teeth and flipped it up and down. Busse looked a little raw around the edges this morning; another night at the Scoreboard Lounge for the vice-president. "I don't really know what I'm doing here, but I'm having an awfully good time doing it," he said. "You'll need a hard hat." He crossed the room and pulled a hard hat from a shelf and handed it to me. I set out to explore the first major greenfield steel mill to be built in the United States in more than twenty years.

■ ■ ■

A half-dozen construction trailers sat next to the Admin Building, where the Run-a-Muckers carried on with the business of building a steel mill. The construction trailers were otherwise known as Texas condominiums. In Trailer No. 4, inside a smoky office jammed with blueprints, a "steel-worker" named William W. Hawley stared at a blueprint, smoking a nearly dead Macanudo cigar. The Macanudo was a five-dollar job.

"Nobody's doing anything today," said Will Hawley. "Everybody's hurting from last night at the Scoreboard." He squeezed his eyes shut and then opened them wide. "Another night at the Scoreboard. I may be the only person at this company who's ever drunk two Nucors without puking."

Will Hawley was a tall fellow in his early thirties, a native of Indiana. He had a fleshy face, straight brown hair going thin on the cowlick, a mustache, and dark-blue restless eyes. He had never seen liquid steel in his life. "I'll be a utility man in the hot mill building," he said. "Kind of like a utility outfielder."

Hawley turned his attention to his cigar. He had smoked it down to a stub that resembled a drum of sushi wrapped in kelp. Hawley squeezed the stub between thumb and index finger and sucked air through it. There was no hope for this Macanudo. He placed it on a heap of Macanudo butts piled in an ashtray.

"I need to run up to the melt shop and talk with Duane Hurler," he said. "You want to have a ride on his crane? He's got this little travel platform on his crane. He'll take it up on a cable, and you can get a view of the Project from up there. He'll probably free fall us. Do you think you can handle that?"

"What do you mean?"

"He'll drop us from the end of his stick."

Hawley explained that Duane Hurler would take an innocent person up to the tip of his crane on the travel platform and then, suddenly, without warning, he'd let the platform fall from the tip of the crane, at the end of the steel cable. The platform, with the victim clinging to it, would fall the height of the crane—a crane drop—and then Hurler would jam on the brakes just before the platform hit the ground. The victim thought he was in the middle of an industrial accident and was going to die, and he would scream horribly and probably blow a sphincter on the way down. In Hawley's words, "I've never been on a crane drop, but I've heard it's a different experience."

In preparation for a visit to Duane Hurler and a possible crane drop, Hawley put on a pair of mirrored Vuarnet Cat's Eyes sunglasses and a green standard-issue Nucor hard hat. Hawley had stuck an embossed tape label to his hard hat that read "Unguided Missile." "You got a hard hat? Good."

I put on my hard hat and followed Will W. Hawley out of Construction Trailer No. 4. He crossed a parking lot on a pair of flexible Hoosier legs and then swerved and headed for a pile of broken boards and aluminum fragments bulldozed into a heap next to the construction trailers. "Take a look at that," he said. "That used to be our blueprint trailer." It was a wrecked Texas condominium, and it looked as if it had been hit by a bomb.

The destruction of the blueprint trailer had been a slight setback for the Crawfordsville Project. Hawley explained what had happened. The trailer had contained all of the master blueprints for the Crawfordsville Project. A week ago, in the middle of the night, a windstorm had come out of

■

nowhere and exploded the print trailer. It burst apart, and the prints went up in a cloud of paper and were scattered in the mud across the job site. Some blueprints were never found, and it seems possible that they were lofted clear out of Montgomery County. It was as if the finger of God had reached down from a cloud and touched Nucor's master designs, and God had said, "This thing will never work," and threw the plans away. But since Nucor was designing the steel mill as it was being built, perhaps not even God knew what it would look like.

We climbed into a pickup truck, Hawley started it with a roar, backed it around, gunned the truck, and lurched across the job site. It was a windy, humid day. A creamy sky, stiff with wind and bearing a threat of afternoon tornadoes, arched over the Project.

■ ■ ■

We found Duane Hurler at the west end of the melt shop, sitting inside the cab of a pickup truck, with the windows rolled up and the motor running at high idle. He rolled down his window and a tendril of cool air greeted us. He was running his air conditioner.

"So what's happening?" said Hawley.

"We're just hangin' joists today," said Hurler, in a thick voice. "I want to hang *iron*."

"How are you feeling?" said Hawley.

The iron hanger's long nose twitched as if he had been unable to clear it of its blockage; it must have been plugged with a snot the size of a pinto bean. He reached behind his seat and pulled forth his little chrome beaker and shook it gently beside his ear, listening to it slosh. "I bought a gallon of this from a guy in Mountain Home, Arkansas," he said. "I have got some people messed up on it." A sly grin traveled over the iron hanger's face and got all involved with his beard. He hid the flask behind the seat, lit a cigarette, and turned his gaze to the activity overhead, which he was supervising. He wore a fiberglass hard hat reversed on his head so that he could look upward without the brim getting in the way. The hard hat had a sticker on it that read *Manitowoc*—the brand name of his crane.

Two parallel rows of concrete piers stretched into the distance, running from east to west—the backbone of the melt shop. The concrete piers of the melt shop were 30 feet high. On the piers stood steel columns that were 90 feet high, and roof joists rested across the columns, spanning the melt shop with voussoirs of structural steel, as if the melt shop was a flat-roofed

■

cathedral. The melt shop was 130 feet high; almost exactly the height of Reims Cathedral.

Red River iron hangers sat on the roof joists, 130 feet off the ground, gripping the joists between their knees. At the south side of the melt shop, on the ground, Red River men with arc welders were welding Nucor joists into roof assemblies, and their arc welders glittered like stars that had fallen to earth.

An enormous Manitowoc erection crane was stationed at the west end of the melt shop. The crane grunted and lifted a joist assembly into the air, turned, and delivered it to the iron hangers on the roof of the melt shop, who wrestled the assembly into place across a span of columns and hammered bolts into it to fix it in place.

Duane Hurler controlled the crane and the iron hangers from his air-conditioned truck, pointing here and there with his finger. Hurler pointed wordlessly at a workman standing on a column, a hundred feet over our heads, who nodded to Hurler and began to whack at a bolt with a hammer. Duane Hurler was building the melt shop with his finger.

Hurler grabbed a bullhorn from the seat and aimed it out the window at an iron hanger who sat on a joist over our heads. He punched the talk trigger and said, *"Clip yourself in."*

The iron hanger fished a nylon strap from his waist and threw the strap around a joist and clipped a ring to the strap.

"That's a wind blowing up there," remarked Hurler, squinting. "They're cooning pretty heavy up there."

The Red River men inched along the joists with their legs wrapped around the joists. "Cooning" is traveling close to the iron, in the way that a raccoon is supposed to travel along a slender tree limb, with its legs wrapped around the limb. The men wore hammers on their belts, and the wind blew the hammers against the steel with violent clangs.

Hurler muttered, "When you don't git it between your head that you can be killed up there."

A Red River boss in a white shirt came over to us. The man had a beard and wore sunglasses of the aviation type. He was the Red River project manager. The boss nodded to us, put his hands in his back pockets, and squinted upward, studying the exquisite craftwork overhead through his sunglasses. Then the boss turned his attention to the ground, to a work crew with a dump truck. The crew was pouring concrete into a plywood form at the bottom of a hole. "Those guys from Superior Concrete are pouring as fast as they can, but they still can't keep up with Duane," said the Red River boss. "I bought Duane that bullhorn. Duane with his bullhorn and that finger of his waving around have put us ahead."

Duane Hurler removed the cigarette from his mouth and spat a fragment of tobacco off his tongue. "We give you your best erection," he commented. He twitched his nose with his thumb and inhaled from his cigarette. "If I don't hang some iron soon, I am going back to Arkansas. I really am. I really am. I am going home to my honey and I am *not* coming back here."

Hurler and his boss were upset with Nucor. Nucor took competitive bids for construction jobs, lump-sum bids, a job for a fixed fee. Red River Steel had been the winning bidder to erect the structural steel for the melt shop. Red River needed to finish the job quickly in order to make a profit on it, but Superior Concrete, the contractor that had won the bid to pour the foundations of the melt shop, was still pouring the foundations. Hurler was waiting on the Superior boys, and every day that the job dragged on, Red River Steel saw its profits dwindle.

Will Hawley said to the Red River boss, "Did you try any of that moonshine last night?"

"Are you kidding?" replied the boss. "Duane's stuff will make you mean. I was drinking whiskey."

Whiskey had not softened the Red River boss's mood; this morning he had a hangover that made his sunglasses bulge from his head like black blisters. "I was throwin' rocks at a Nucor guy this morning," he remarked. "I threw a couple a rocks at him and told him to get out of the melt shop before I got into a fight with him. It wasn't a fight yet, but it was heading that way. It was for his own good. It was a safety measure."

"You need to improve your attitude," said Hawley.

"I don't believe these Nucor sons of bitches can feed me enough whiskey to improve my attitude."

"Those columns, I set them all," remarked Duane Hurler, waving his cigarette at the ninety-foot steel columns that sat on the thirty-foot concrete piers. To hang a column, he took the controls of the crane himself. "I pick up a column, I highball the crane, I give it all it's got," he said. "I floorboard it, and I walk him back, and that column, he steadies comin' on up, and I hang that iron in one piece."

Hurler pointed with his cigarette toward a cluster of massive concrete piers near the east end of the melt shop. "They're gonna put the casting machine over there," he went on. "You see how those piers are concrete? If molten steel gets loose around there and those piers were made of steel, it would be liable to melt this place down."

■

Will Hawley nodded thoughtfully, and said, "Yeah."

A big spill can wreck a melt shop and kill steelworkers like flies. Hindsight has a way of changing a story, but I sometimes wonder if I'd had a feeling that death was warping toward the melt shop.

"How much stick have you got up there?" asked Will Hawley, indicating the crane.

"That's two hundred and forty feet of stick," answered Hurler. "With a jib boom you got two hundred and eighty feet up to the end of that crane."

"That's your Manitowoc 4100 W," broke in the Red River boss. "Your Manitowoc is the Cadillac of erection cranes." He pronounced the word *Manitowoc* as "Man-a-Walk." "It's a seventy-five-ton Man-a-Walk, but it'll pick up two hundred tons if you put a ring around the base for stability. You try to pick up two hundred tons without a ring on that crane, and that Man-a-Walk will turn over. There is bigger Man-a-Walks than that. The biggest Man-a-Walk will pick up a million pounds. You can pick up a nuclear reactor vessel with that crane and walk the vessel right into a containment building. I've seen it done." The boss remarked that Red River was doing a supergood job on the melt shop, unlike some of these other steel erection firms around here, who hung their iron all crooked. The boss strode away, scuffing at dirt, his sunglasses protruding from his head.

Hawley turned to Hurler. "We came to see if you'd give us a ride on that crane," he said.

"A *ride*, huh." Hurler stroked his cigarette hand across his beard and a row of white teeth materialized in the beard. "You bring a clean pair of shorts with you? You're going to need them afterward."

The Manitowoc, the Cadillac of erection cranes, with a total of 280 feet of stick, stood half the height of the Washington Monument. The crane had a two-part boom, consisting of a main boom, 240 feet high, used for lifting heavy loads, and a jib boom, a slender truss 40 feet long that stuck out at an angle from the top of the main boom like a bent fingertip. A single cable ran all the way up the stick to the tip of the jib boom: This was the crane's whipline. A ball and hook dangled from the whipline.

Hurler punched the trigger on his bullhorn and waved his finger. "*Hey. Drop that whipline.*" His voice echoed over the job site.

The whipline paid out from the tip of the jib boom, and the hook raced toward the ground. The hook stopped abruptly, 3 feet off the ground.

"*Pick up that cage and bring it here.*"

The whipline twitched around, delivering a small travel platform to our feet. The travel platform was the size of a coffee table and was made of plate steel. A railing extended around the platform, waist-high. There was no other protection on the platform.

"Wait a minute," said Hawley. He walked around a pier.

Duane Hurler kept his eyes on his men overhead. The back of his neck was creased with black grooves where dust had settled in cracks. Suddenly he remarked, "I dropped a guy last week. And. He. Like. To. Have. Lost. His. Ass."

Will Hawley returned. "There's a Port-a-Pot around the corner," he said to me. "I advise you to empty your bladder before we go up."

I took Hawley's advice, and then Hawley and I stepped onto the coffee table at the end of the whipline. A tanned, muscular iron hanger from Texas, who had a slight smile on his face, joined us.

Duane Hurler gave us some safety instructions. "When they free fall you, you'll hear the whipline singing in the winch. And then you will like to lose your ass! You hang onto that rail! You hang on, 'cause your hearts will flutter! *Oooooo,* that will make your hearts flutter!"

"I want you to ride with us, Duane," said Hawley.

Hurler looked appalled. "You do not want that. If I went with you on a crane drop, my boys would try to kill me when they dropped us." He made a signal with one finger, and the whipline tightened, and the Manitowoc grunted and suddenly lifted us 130 feet past the iron hangers on the roof of the melt shop.

They cooned in the wind, their feet locked beneath them. They were watching us. One guy wore a red handkerchief tied around his head, and the handkerchief fluttered in the wind. Their eyes followed us upward as we traveled another 130 feet or so, until we stopped at the tip of the jib boom. Now the iron hangers on the roof of the melt shop were tiny figures far below us, and the red handkerchief was a droplet of blood clinging to a length of steel.

The wind was strong in the upper air. It buffeted and tore at the travel platform as we dangled over the melt shop. The platform pitched and rolled and heaved like a boat riding at anchor in a gale, and the crane's jib boom undulated like a mooring cable anchored to the sea floor.

The horizon tilted this way and that, and the Project rocked back and forth. It spread out in all directions at our feet: the crystallization of capital. Directly below us stood the melt shop, with its row of columns lined up on an east-west axis. North of the melt shop stood the hot mill, a building lined up on a north-south axis at right angles to the melt shop. The melt

■

shop and the hot mill formed an L. At the angle of the L there was a cluster of piers: the roots of the casting tower. The Compact Strip Production machine would start at the casting tower and run from south to north in a straight line, springing from the casting tower at the east end of the melt shop, going northward into the hot mill building. The hot mill was half-covered with a steel roof.

"They haven't even finished designing the other half of the hot mill," said Hawley.

In the center of the hot mill there was a hole in the ground large enough to serve as the root of a Manhattan office building. The hole marked the site of the CSP's millstands, rolling machines that would flatten the red-hot emergent slab of steel down to the thickness of rawhide. The cold mill building stood alone, northeast of the hot mill, and it was now covered with a roof.

We could see for twenty miles. White barns and white silos stood here and there on the landscape. To the south we could see Whitesville, consisting of a few houses and one grain elevator. The old Monon tracks, now a Conrail line, passed through Whitesville, angling toward Chicago; the Monon was one of two railroads that would haul the Rust Belt to the Crawfordsville Project. Far to the north, in the whitened air, a line of clouds popped up through haze: the Lake Shore, where the Big Steel mills were making steel. Road graders and dump trucks crisscrossed the job site, trailing streamers of yellow dust. The wind dragged the dust through the bony buildings into plowed land, where the dust unfurled and dissipated.

"We're going to be Number One in steel," said Will Hawley.

"You seen enough?" said the Texas iron hanger, in a bored voice.

"I'm not sure," said Hawley, nervously.

The iron hanger pulled a walkie-talkie from his pocket and said to it, "Drop us."

Two people screamed, neither of them the Texan. The platform fell away from our feet. Our feet extended on tiptoe in zero gravity and we gripped the platform's rail, staring down at Hurler's tiny truck and at some specks near it, which were iron hangers on the ground, watching our high dive. It was a bungee jump at the end of a rigid steel cable. We fell half the height of the Washington Monument while the whipline sang in the winch. I was afraid that my feet would drift off the platform and kick over my head. It was like being trapped in an elevator that has broken loose inside a thirty-story building. You float toward the ceiling, the hair stands away from your head, and the contents of your stomach bulge into your

■

esophagus, and then you wait for the obliterative impact. We floated past the iron hangers on the roof of the melt shop, and their faces tracked us downward.

We came to earth in a severe bump. The brakes jammed on and the platform stopped three feet above the ground, driving us to our knees. "You all take it easy, now," said the Texan. He hooked a leg over the rail and walked away.

Hawley and I stood up and climbed slowly off the coffee table. Hawley's face showed no expression behind his mirrored Vuarnets, but it was as pale and moist as a turkey loaf. "Pretty unusual," he said, and coughed once.

"Don't that make your heart flutter?" said Hurler, leaning against his truck. "*Oooooo*. Come here." He reached through the window and pulled out the chrome flask. "It will help the fluttering," he said.

I took a gulp of moonshine, another gulp, and handed the flask to Will Hawley, who upended it and drained it, pulled the flask off his lips with a pop, and handed it back to Hurler.

"My crane operator cut you some slack," said Hurler. He peered into the empty flask. "If I had been at the controls, your hearts would have *stopped*. I tell you what. If anybody asks you what happened, you say my boys free falled you according to OSHA standards."

■ ■ ■

F. Kenneth Iverson and Keith Earl Busse are driving around the Crawfordsville Project in a pickup truck, wearing green Nucor hard hats. Busse is behind the wheel, pointing out progress to Iverson. Iverson grips the dashboard and when the truck lurches he slides this way or that across the seat.

"A lot of people are saying the casting machine won't work," says Busse, "but I've seen the holes in the hands."

"I just *know* something won't work," says Iverson. "But there's enough talent around here to fix it."

"Our people are not sure what the beast is going to look like," says Busse.

"I'm not worried about the technology," says Iverson, hanging onto the dashboard and peering out the window. "What I'm worried about is the cost of starting up this mill and the amount of time the startup could take."

Busse drives the truck inside the cold mill building, the separate building that will contain cold-rolling machines. The cold mill is a cavernous

■

space with unfinished walls open to the weather. Busse parks the truck. He and the chairman climb out, shake hands with a few steelworkers, and descend a flight of stairs into a tunnel. They splash around in water at the bottom of the tunnel, Busse pointing out pipes to the chairman.

Meanwhile, something like the following thought is going through the chairman's mind: *One of the things these managers like to do is take you down in tunnels when they're building a steel mill, to show you what they've done. I don't care if I ever go down in another tunnel.*

They climb a flight of stairs out of the tunnel to ground level and they get back into the pickup truck, and the truck leaps out of the cold mill into open country. Busse stops the truck with a jerk in a field of mud near the cold mill.

Four enormous steel objects are lying flat on railroad ties, floating in the mud. The steel objects are shaped like elongated doughnuts. The doughnuts are thirty feet long and ten feet across, and they are made of forged steel; they are incredibly massive lumps of solid metal. They look like chain links. They are known as roll housings, and they will be placed upright inside the cold mill, mounted in concrete. Roll housings are the skeletons of a steel-rolling machine. Roll housings are chassis structures that hold and support rolls. Rolls are enormous cylinders of cast iron, like rolling pins, and they roll steel flat.

Iverson wants to inspect a roll housing. He climbs down from the pickup truck. Then he puts his foot up on a railroad tie and heaves himself onto the roll housing. He stands up on the forged steel, looking around. A gust of wind plucks at his necktie. Iverson is wearing pin-striped trousers stuffed into a pair of steel-toed boots. He kneels and runs his hand lightly over the steel.

Busse says, over a rising wind, "Every day more crates of this stuff arrive."

Iverson grunts and straightens up and touches the brim of his hard hat, looking west. "How far does our land go?"

"All the way to those trees on the horizon." Busse's thoughts run like this: *Room for more manufacturing facilities.*

Spikes and hummocks of stormboil pile along the horizon. The clouds look like popcorn. An anvil cloud has flattened against the tropopause, deploying feathery edges. Sunlight splashes across the plains, nailing a silo with a burst of illumination. The anvil cloud is putting a shaft of rain into bare fields. Puddles of water all over the job site reflect the sky.

"Everything is up to us now," says Iverson, with his hands in his

■

pockets, scanning the holes that his company has dug in Indiana. According to Iverson's later recollection, something like the following thoughts are going through his mind: *This job site is in terrible condition. I've had experience before with job sites in spring. We are already behind schedule. The mud this winter was very bad. With mud like that, you are stuck for three months and you can't do a thing, you can't even move the earth-moving equipment, all you can do is pump water out of the holes you've dug in the ground.* The question is the startup. The startup is now only one year away. It's the question of what will happen when hot metal is poured into the machine. The young people are going to learn about steel. Iverson's mind is working ahead a year in time, toward the startup. The company will have to push the startup. By his own later account he is thinking, *When can we get this thing started up? We have set a date to get hot metal out of a furnace by March or April of next year, but I don't know whether we can make that date. We are shooting to get hot metal out of a furnace one year from now, and all we have here is holes in the ground and some structural steel.*

■ ■ ■

At the restaurant in the Holiday Inn in Crawfordsville, Iverson and Busse talk about the looming war with Big Steel. They have had a cocktail or two in the lounge. Busse has been drinking his usual Canadian Club and Seven-Up, and Iverson has been drinking his usual vodka on the rocks, with a slice of lime and never, ever any vermouth. Now they're eating dinner. Busse is cutting into a sirloin steak and Iverson is working on a piece of unidentifiable fish. The traveler who orders seafood in the Midwest ought to have his head examined, but Iverson orders fish everywhere he goes.

Busse is hunched over his sirloin, talking in his argumentative way. He is telling Iverson that he is worried about the future of the Nucor Corporation. He is telling Iverson that he is afraid that Nucor could someday resemble Big Steel, and that would be the kiss of death, in Keith Busse's opinion. "How do we allow Nucor to grow without expanding the bureaucracy, Ken? When I go up to visit the Lake Shore to Big Steel, I see vice-presidents stacked on vice-presidents, research departments, assistants to assistants, all kinds of weird-ass relationships that I can't even understand."

Iverson nods and cuts into his fish. "Yeah," he says, munching it. "They have in-house consultants to help them design a plant. Then after

■

they've built it they hire an outside consultant to tell them if they built it right.''

"All those Big Steel companies are the same," Busse goes on. "They are a culture, it's a bureaucratic culture, and there isn't a nickel's worth of difference between 'em. But, I mean, Nucor is getting so big. We may need another layer of management to carry us into the next century." An idea lights up Busse's face. It's time for a Negotiation with the chairman. Busse says, "Maybe we're gonna need group vice-presidents."

Iverson's fork pauses over his fish. "*Group vice-presidents!* Keith, do you want to ruin this company?" He lowers his voice. "That's the one thing I'm dead against, group vice-presidents. Because then you become a bureaucracy." A sly grin crosses Iverson's face. "That's the same old Harvard Business School thinking, Keith."

Keith Earl Busse has never been near the Harvard Business School in his life, and he takes the remark badly. "Well, all right then, Jesus Christ—be creative, Ken!"

"Be creative? God damn it, Keith, I want us to be more creative than that—"

"We've got to the point where we are so damn big—"

"God damn it, Keith! We don't need any damned group vice-presidents! God damn it!"

People at other tables are starting to notice.

The chairman lowers his voice. "Because it removes the management even farther from the employees. And you're just putting somebody in there, that's all you're doing, Keith. There's no point to it."

Busse won't give up. "I'm really worried about the span of control. How many people can report to one person before that one person loses control, Ken? We've got so many plants out there, we're at the point where you're trying to manage a large number of hard-charging managers with big egos and swelled heads! The span of control is too large!"

Iverson agrees that Busse has a point. There are certainly a large number of swelled heads among Nucor's middle managers, Busse's head being one of them. But he tells Busse that Busse's idea for group vice-presidents is a rotten idea.

Busse makes the following point. He suggests to Iverson that it would be possible to take some of the general managers of Nucor's factories and promote them to group vice-presidents, to give them responsibility for several factories. "That way, you're only creating half a new management layer."

■

Iverson's gray eyes go wide. "*Half* a management layer! I don't buy that one bit!"

"Ken, don't give me that shit!" says Busse. And Busse is thinking, *Uh-oh, I'm talking to the chairman of the board, here.* He keeps going. "For ten years you've been saying we're going to give away the candy store and we haven't done it yet."

"We don't need any group vice-presidents!" says Iverson. "All they do is get in the way! Because you have to include them in decisions!" Iverson is getting steamed, but he has a little rule of thumb, that a good businessman is hard to bruise and quick to heal, and at a moment like this he tries to remember his rule. "We could at least double our size before we really even need to think about a new layer of management. We could go to *two* billion dollars in sales without putting a strain on the staff."

Iverson has been saying this for years. For years he has been telling financial analysts and business professors and his own managers that Nucor could double its sales without putting a strain on the corporate staff. In the meantime Nucor has doubled its sales, and doubled them again, and doubled them again, and doubled them again, and doubled them again—five doublings, all told, exponential growth—and Iverson is still telling the analysts and professors that Nucor could double its sales without putting a strain on the corporate staff. Nucor has mushroomed in splendor, wealth, and technical prowess, the Cinderella of the American steel industry, but the headquarters is still located in a rented office beside a shopping center, just where it was twenty-five years ago, when the Cotswold Building was modern and attractive and didn't have that glazed-vomit look; and the corporate staff has grown only marginally in all those years.

Keith Busse feels the pressure of a brilliant future hanging over the company, the possibility that a few CSP machines could grow Nucor rapidly into one of the largest steel companies in the United States.

Iverson thinks that Busse worries too much. He also thinks that Busse argues with him merely for the sake of an argument, and that once Busse gets into an argument started by Busse, Busse won't back down. In Iverson's words, "Keith is a tiger. He will get hold of something and he won't let go of it."

Iverson won't let go of it, either. "Keith, I hope that by the time we have group vice-presidents I'll be collecting Social Security!"

■ ■ ■

■

Iverson stays that night in the Crawfordsville Holiday Inn, in a single room ($46) looking out on a feed hopper. The next morning, he drives his Hertz Lincoln Town Car at a fast clip to the Indianapolis airport, pressing his advantage in traffic, hugging the right shoulder of the road. Iverson likes to rent the heavy iron when he travels, your Town Cars and your Cadillac Sedan de Villes. It's his one indulgence, because a luxury car in a wreck gives you that edge for survival. The car turns into an armored barge of fender metal: wildly out of control, to be sure, but it will burst you through a Honda without straining your ribs.

He flies alone back to the city of Charlotte. He flies tourist class, and on a Supersaver ticket when he can get one. He tries to build up his frequent-flyer mileage, so that when he travels with his wife, Martha, they can get a free upgrade to first-class seats. Martha refuses to fly tourist class with her husband, and her reasoning goes something like this: Ken, I don't like those narrow seats, and for heaven's sake, people with our kind of money don't always have to fly tourist class. So he saves his free upgrades for Martha. Top executives of other Fortune 500 companies, anxious to meet Ken Iverson, eager to learn a few management tips from the chairman of the extraordinary Nucor Corporation, sometimes offer Iverson a free ride on a corporate jet if they are heading to the same destination as Iverson. But he refuses to fly on corporate jets, because he doesn't want Nucor's steelworkers to see him stepping off a corporate jet. Flying tourist class on steel business, he has to deal with the occasional crying baby or the lonely traveler in sales, but he does paperwork during the flight, so most people leave him alone. Nobody recognizes him for being the most carefully watched executive in the world steel industry. "Since I travel alone," as he explained it, "it would be an absolute waste of money for this company to have a corporate jet."

Martha and Ken Iverson have consoled their earthly existence with a fine house. Their children have grown up and married—they have grandchildren, now—and like many couples whose children have left home, their house seems too big for them. It sprawls in a series of glassed pavilions along the shore of a small lake in Charlotte.

An African parrot in a cage in the front hall greets him in his wife's voice when he comes home: "Hello, Ken! Where are you, Ken?"

"Hello, Caspar!" he replies to the bird, and to Martha he shouts, "Hello, I'm home!" Then, after he and Martha have caught up on whatever has to be caught up on, he goes outside to the lake to hack around with his birds.

■

Iverson keeps four black swans, eight Chinese geese, six mandarin ducks from Korea, and uncounted flocks of wood ducks, mallards, and Canada geese. His birds number as many as a hundred, depending on seasonal migrations. Iverson has noticed that the same individual Canada geese drop in on him year after year. His exotic birds have pinioned wings, and so they remain on the lake all year.

He walks down to the lake carrying a fifty-pound sack of meal on his shoulder, food for his birds, down a flight of brick steps, on swinging legs. He regards his walk to the lake with a sack of meal across his neck as a weightlifting routine, a clean and jerk of swan food.

He stands on a graveled shore. "*Hullooo! Huh-hoooo!*" he shouts. His favorite bird is a black swan named Jessica. "Jessica! Where's my Jessy? *Wah-hoooo!*"

The black swans paddle and flap toward Iverson, honking. They are powerful creatures, with a six-foot wingspan. A black swan can break a person's arm with a single blow of its wing. Iverson's swans have given him some wicked bruises, but he doesn't mind the bruises, because a good businessman is hard to bruise and quick to heal. Dogwoods are shining in the woods. The black swans move toward Iverson across the mirrored lake on a May evening like carbon smoke drifting over a heat of liquid steel, notifying him of something exquisite in nature, something innocent and unexpectedly harsh. "*Wha-hoo! Wha-hoo!*" he calls, stripping the bag and dumping it into a galvanized steel feeding trough, and the swans gather around him with beating sable wings.

4

RUN-A-MUCKERS

■ ■ ■

Keith Earl Busse's gun store in Fort Wayne, Indiana, sells guns under the sign of H & H Firearms. Busse describes H & H Firearms as "the prettiest little gun store in five states." It is a low structure made of split cinderblocks, almost without windows, situated on Coliseum Boulevard, two miles west of the grave of Johnny Appleseed. In certain respects it is a gun supermarket. H & H Firearms is the sort of place that gives those who are against guns a feeling that the world is coming to an end, while it gives those who are in favor of guns a desire to shop.

The only windows that pierce the gun store are in the gun display area, facing away from traffic. Busse and his business partner, a stout, genial gun-toter by the name of Bill Hartsing, did not care to install too many windows because a lot of shooting goes on inside the walls. Busse has a class III federal license to buy, sell, and transport fully automatic weapons, and he is doing a brisk business with judges.

"These judges in Indiana own guns," Busse remarked, as we wandered among display cases offering expensive heaters for sale, gleaming with chrome and blued steel. "You wouldn't believe what some of these judges own. Every now and then some judge dies, and Bill, my partner, goes out and gets us an unbelievable collection. Uzis, Thompsons, .50-caliber stuff on a tripod." Checks and balances.

"Our customers are not your backyard woodchuck," said Busse.

We stopped at a display rack that held plastic bags of gun wadding. "This is your black-powder stuff," he explained. "Over here is your

■

police hardware. Here's your Uncle Mike's line of holsters. Here are your concealable holsters.'' He pointed to a display case. "Here are your Smith & Wesson .44 Magnums,'' he said. "Our main long guns are the Winchesters. We've got probably eight hundred guns on display here.''

On a good weekend, at one of the regular gun shows at the Coliseum in Fort Wayne, Busse and his partner could move twenty-five thousand dollars' worth of rods onto Hoosiers.

High on the walls, out of reach of tire-kickers, Busse and his partner had hung the black equipment. "Here's the infamous Thompson submachine gun,'' he said, pointing to the type of gun that Ernest Hemingway had once used to machine-gun sharks off Key West. "Here's an Uzi, here's a Schmeisser machine pistol. Here's a Colt AR-15, paratroop model. Here's an FBI-type .00 semiautomatic assault shotgun—they used 'em in Vietnam. A shotgun like that'd stop anything in the underbrush, you couldn't miss. You notice the baby Browning up there on the wall? That's an exact replica of a .30 caliber air-cooled Browning machine gun on a tripod. It fires .22 bullets in a belt—see the little teeny ammo cans?''

The salespeople at H & H Firearms, men and women, wore handguns in exposed holsters. They were friendly and courteous. Busse told me that they were certified shooting instructors, taught classes in gun safety, and liked to throw around technical terms like "hydroshock,'' which describes the effect of a bullet passing through a jelly, such as a person. "That's a word they use,'' he said.

We went into an office where Busse had laid out some handguns on a desk. We put on hearing protectors and safety glasses and carried the guns into a pistol range, one of three firing ranges inside H & H Firearms. In a firing booth, Busse fitted a paper torso target into a clip, hit a button, and the target was reeled down the range. He fitted six bullets into the cylinder of a Colt Python .357 Magnum revolver and took a two-handed police stance. *Kak-kak-kak-kak-kak-kak!* He emptied it suddenly, the Python bucking against his wrists. The gunfire shocked my ears, even muffled by hearing protectors.

"That Python gives off a real crack,'' he said, in an interested tone of voice, easing from the stance and handing me the Python through a haze of smoke.

My ears ringing, I loaded six Magnum rounds into the Python, while Busse sent a fresh target downrange. I blew six random holes in the target, the gun's barrel jerking around. Busse hit a button and the target was reeled back up the range.

■

Busse examined the target through his safety glasses, which gave him an academic appearance. "That's a decent cluster," he said thoughtfully.

Next Busse handed me a Ruger Red Hawk .44 Magnum revolver. It was not exactly a concealable handgun. It was as heavy as a spanner. If you tried to hide this .44 under a business suit it would tug one shoulder lower than the other and push the lapels out. This gat would put a bulge in a Santa Claus suit.

"Go ahead, pop the cylinder," said Busse, handing me a box of .44 shells.

With slippery, fumbling fingers, I opened the gun's cylinder and fitted six shells the size of headbolts into the cylinder. Meanwhile Busse gave me a lecture on the kinetic energy of the .44 Magnum round.

"You'll not believe this," he said, "but a .357 Magnum round does not have complete stopping power. The police have discovered that they can put five or six .357 rounds in a guy, and if he's determined he'll keep coming, even with severe exit wounds. But the .44 round has complete stopping power. One shot from this .44 will lift a person right off his feet. Talk about hydroshock."

I asked him if the .44 would kick very much.

"Don't worry about it. It won't shatter your elbow."

I cocked the .44 and pulled the trigger cautiously. The gun gave off a blinding muzzle flash and a boom like a naval salute. Clouds of black, gusty smoke swirled out of the firing booth. The gun began to shake in my hand, the barrel dancing around.

We fired a Browning nine-millimeter semiautomatic pistol with a gold-plated trigger. We fired a stainless steel Colt .45. "Do high-speed fire," he suggested, so I fanned the stainless Colt, but the last round jammed and the gun's slide didn't come back. Busse wasn't happy with it for jamming during rapid fire. That could cost you your life in a tight spot—say, if someone tried to put a hit on you in a parking lot. He fiddled with the gun's slide, aimed the gun deliberately at an angle at the wall downrange, and pulled the trigger with a bang. A little skid mark appeared on the wall.

Busse returned to the display area of the shop, where he collected an ultralight version of the U.S. Army's M-16 rifle. It was a shortened, fully automatic Colt AR-15, an uncommon gun, with an extensible buttstock and a snubbed barrel with a flash suppressor, a gun used by paratroops and anti-terrorist teams. He introduced the gun to me under the name of "Stubby, the Magic Dragon." We went into a heavy-duty firing range, reinforced for supersonic weapons.

■

Busse entered a firing booth and slapped a clip of twenty 5.56-millimeter NATO rounds into Stubby. There was a customer in a firing booth near us, shooting a deer rifle. Busse called out to him: "We're gonna make a racket over here, so don't be alarmed." He beckoned me into the firing booth and handed me the device. "Pull the stock out," he said. I pulled out the stock, which locked into place with a click. "You know what the 'feed' on a machine gun is?" he said.

"No."

"Here's the feed," he said, reaching in front of my face and snapping a T-shaped metal fingergrip. That delivered the first of the twenty NATO rounds into the firing chamber; it was like cocking a pistol. "Set Stubby on 'full auto,' " he said, pointing to the gun's fire-control selector, a switch on the side of the gun.

I set the fire-control selector on "full auto."

"Now look," said Busse. "Stubby's gonna go off fast. You've got to pull *down* on the barrel, because any machine gun will want to kick up on you. Obviously I don't want you to fire any bullets into my ceiling."

I touched the trigger and a roaring welter of kicks almost backed me out of the firing booth. The Dragon emptied its clip in less than two seconds, NATO rounds whanging across the floor and walls, but I'm confident that not more than half of the bullets hit the ceiling. A wicked cloud of smoke billowed downrange. The customer with the deer rifle stared at us.

Busse reeled in the paper target and inspected it. "You missed," he said.

■ ■ ■

Busse believed in gun control of a certain kind. He believed that nickel-and-dime gun dealers should be brought under control. He thought that the price of a gun dealer's license ought to be raised sharply, in order to put the trade into the hands of smooth professionals running gun supermarkets, as he did, offering a wide variety of products to the consumer on razor-thin margins. He thought that the government should crack down on small gun dealers, whom he claimed might sell a Kalashnikov to a wing nut without asking too many questions. "You need to get rid of some of the chaff, weed out the barnyard ducks," as he expressed it. "The National Rifle Association takes the position that any bozo can sell guns out of his basement. You can't sell liquor out of your basement." He thought that there should be a computer search of the psychiatric records of any person who wanted to buy a machine gun. The wing nuts ought to be distinguishable with good software.

■

Nobody at Nucor seemed to worry about Keith Busse and his machine gun business. Possibly the guns amused Charlotte. ("Keith's gun shop probably doesn't take up as much time as golf, now, does it?" remarked a certain Hugh David Aycock, Nucor's president and chief operating officer.) More than driving a golf ball down a fairway, Keith Busse enjoyed a demonstration of firepower.

He liked fast cars and fast snowmobiles. He had owned a collection of Corvettes, but got a little tired of them and switched to Porsches. He liked to goose a snowmobile through the woods at high speed—now there was a good way to experience nature, with the trees flashing past and your lips dragged back and nothing in your ears but the howl of internal combustion. He liked to go to some lonely place on a farm and blow up something with a machine gun, looking for interesting effects. A target might be a rotten stump, an arrangement of bottles and cans, a telephone book, a power-cable spool, an abandoned house, a watermelon. He liked the sensation of action at a distance, of things falling apart and being chewed up when you touched a button. He wasn't much of a hunter, it seemed boring and a little cruel to pump a couple of bullets into an animal, but to hit an abandoned house with overwhelming firepower, to knock down the brick chimney with an assault shotgun, to watch the chimney explode, *ker-bash,* in a shower of bricks sliding down the roof, and then to take up a machine gun and blow out windows and walls on full auto, throwing glass and wood fragments through boiling clouds of plaster dust, now there was a hobby that took your mind off business, and certainly it relaxed him more than chipping a golf ball out of a sand trap. He called the art and craft of blowing up an object with a weapon "plinking," and he could plink as long as his ammunition held out and he had an interesting specimen of junk in his gunsights.

A modern steel mill could deliver you the plinking effect in screaming, heart-stopping abundance. Talk about firepower, consider the firepower packed into a Compact Strip Production facility. He likened the CSP to an ultracaliber automatic weapon, "a million-ton shooter," as he phrased it. Talk about hydroshock, the CSP could lift Big Steel off its feet. Keith Busse liked machines. He especially liked powerful machines. Machines were cyclical devices; they operated with reciprocating, repetitive motions, *kak, kak, kak, kak.* Machines extended the human reach into unexplored realms of energy. He especially liked machines that operated at blinding speed with hard sleek fierce involuted movements and with absolute precision while offering a sensation of fantastic violence, and that was what

■

a modern steel mill was all about. "I never dreamed I'd be running a steel mill," he said. It pleased him to think of hot steel roaring out of the CSP, with the machine's fire-control selector set on "full auto" and its muzzle pointed at the watermelon of Big Steel.

■ ■ ■

The Lake Shore extends south by east from Chicago on Interstate 90, along the curving lower belly of Lake Michigan, encompassing the northwest corner of Indiana. The Lake Shore is now the core of the American steel industry, the bedding ground of the companies that survived the late wreck of American steel. On the Lake Shore are parked the flagship American steel mills, lined up like ducks in a row. The flagship American steel mills sit right next to each other, between Interstate 90 and the Lake Shore, along a waterline indented with harbors. One day I drove out of Chicago along the Lake Shore. The air wasn't bad, because the blast furnaces now recycle their gases and clean their smoke. Near the Chicago city limits I came to the United States Steel South Works, sitting against the lake, and back from the lake there was the Acme Steel Corporation; then, crossing into Indiana, I came to a complex of mills belonging to the LTV Corporation, operating in bankruptcy; then to Inland Steel Industries, Inc.; then to the gigantic United States Steel Gary Works, the flagship steel mill of the USX Corporation. I stopped at the Gary Works, but the mills go on: There's National Steel; and there's the massive Burns Harbor facility of Bethlehem Steel, a heavyweight mill at the end of the line. The Bethlehem Burns Harbor mill was the last great steel mill to be built in America, completed in 1967.

The Gary Works, a unionized steel mill, reaches for six miles along the Lake Shore, is a mile and a half deep, and is covered with mill buildings. The Gary Works is a city of horizontal skyscrapers, the city of Troy in the wars of steel. John Goodwin, the general manager of the Gary Works, operates out of a turn-of-the-century brick building not far from the entrance to the Gary Works. Goodwin is widely admired inside the steel industry. Goodwin is an operations man, a manufacturing wizard, reputed to be one of the best in the American steel industry. He reports to the USS division of the USX Corporation, which occupies the USX Tower in Pittsburgh. Goodwin has been to hell and back in the wars of steel, and he is not easily intimidated by Nucor or anything else. He is in his mid-forties, a stocky, pleasant man with a nubby mustache and heavy eyebrows and a

■

shrewd alertness on his face, and he wears a stone watch and a gold ring as big as a hog vertebra.

"At one time I was the manager of the rod mill at the U.S. Steel Fairless Works, near Trenton," said Goodwin. "We had a brand-new rod mill. Then Raritan River Steel, a minimill, came on line in New Jersey with a minimill for making steel rod. We took the stance that the minimills would never be able to match the quality of our rod steel. We were wrong. You know where that Fairless rod mill is now? It's in China. We closed that mill and sold off those rod machines to the Chinese for next to nothing. They took our machines away, piece by piece."

Goodwin intended to show Nucor a fight. "I underestimated the minimills once and I'll never make that mistake again. But I am not afraid of Nucor. I've got to give them a lot of credit. It took a lot of balls to do what they are doing there in Crawfordsville. Keith Busse doesn't have a steel background, of course, but he does have a startup background—Busse knows startups, he knows what they're like. Sometimes I wonder: Will this mill of theirs be another Wankel rotary engine? You remember the Wankel engine? It was a great idea, but it never really caught on. Time will reveal the truth."

■ ■ ■

The sixty-first floor of the USX Tower in Pittsburgh: the office of Thomas C. Graham, the president of U.S. Steel. A line of huge windows looks south across the Monongahela River through blue air, because most of the steel mills of Pittsburgh are no more.

Tom Graham is a short, balding man in his sixties, with a puckery smile, a physically quiet man who hardly moves, sitting in a black leather chair beside a coffee table. He has a pouch under his chin and steel-rimmed spectacles and rather small hands, which rest carefully and symmetrically on the armrests of his chair. He reminds me of a family physician. He is said to have the respect of the unionized steelworkers in his company.

"Nucor has consciously and skillfully worked the analysts," he says. "But Nucor is a lot more like us than Nucor would like to admit. We tore three layers of management out of this company after I became president. Welcome to the club, Nucor." One side of Graham's mouth puckers in a smile, and his steel-rimmed glasses reflect the blue sky.

In 1983, David Roderick, then the chairman of USX, brought in Tom Graham to save U.S. Steel from what looked like bankruptcy. Graham closed steel mills left and right. He stripped fourteen million tons of annual capacity out of U.S. Steel. Fourteen Crawfordsville Projects, cutting the lard out of Big Steel. At the same time he invested enormous sums of money in continuous-casting machines, trying to get U.S. Steel back up to international standards of efficiency in the making of steel. When he became president, the company needed eleven man-hours of labor to make a ton of steel; now U.S. Steel makes steel with about four man-hours per ton, and that is better than in Japan. Tom Graham nearly tripled U.S. Steel's labor efficiency. He did it partly with the help of so-called contract labor, outside workers who work alongside unionized steelworkers.

"You know, there are all these myths about the steel industry," says Graham. "That we're dinosaurs. That we're lousy managers. That we're not technically progressive. But when Lee Iacocca goes on TV and says, 'We're offering a seven-year warranty on the body of every Chrysler car,' nobody credits the American steel industry. That's American steel in that Chrysler car. That steel doesn't rust, and that's why Lee Iacocca can make his guarantees."

"What went wrong?"

"This is all in hindsight, but we let our labor costs get away from us," he replies. "We didn't pay attention to foreign competition. We didn't pay enough attention to the marketplace. Our biggest failure was in letting the, well . . ." His voice trails off. The word "minimill" enters my mind but not Graham's office.

"What about the idea that you didn't invest in new technology?"

"No!" he says. "I don't put much in that theory. The American steel industry is not homogeneous. That's a red herring."

"Would you ever build a Compact Strip Production plant?"

"It's a free country," says Graham, with a little smile.

■ ■ ■

"We had nine hundred applicants for thirty jobs," said Will W. Hawley, swinging the wheel of a Nucor pickup truck onto a county road, down a line of telephone poles that headed for the horizon. "I used to live in Crested Butte, Colorado," he said. "I had a construction business and a gun shop. I was buildin' condos and sellin' guns." When the Rocky Mountain condominium market toppled over and crashed, Hawley liquidated his construction business, sold his gun business, and went home to Indiana, looking for work, where the ghostly apparition of a steel mill practically in his back yard had seemed like a piece of supernatural luck.

■

Nucor had put Hawley through a psychological test devised by a Nucor psychologist, one John Seres, out of Chicago. Nucor's psychological test, which Nucor gave to all candidates for jobs at the Crawfordsville Project, was supposed to identify goal-oriented people, self-reliant people: Nucor material. The test also weeded out applicants who might sympathize with labor unions. Hawley passed Nucor's head examination and Busse hired him to be a Run-a-Mucker.

Hawley parked the truck in his driveway. He lived in a small house on his family's land, a four-hundred-acre parcel of rolling woods and corn-fields in Ladoga, a few miles south of the Crawfordsville Project. He and his wife had built the house themselves. Hawley's ancestors had occupied that piece of land since 1831, and had lived in Montgomery County since 1824, when the first white settlers were coming into the country. At that time the area around the Nucor steel mill was the Big Woods.

Hawley opened the door of his house and a couple of smelly golden retrievers leaped at him, woggling spit-threads from their lips. "O.K., yeah, O.K.," he said to his dogs. A set of stairs ran up to a second floor. "You see those stairs? They're black walnut. We timbered it here on our land." He wandered down to a creek, his dogs playing around him. "I've caught a bunch of catfish in there," he said, pointing to a deep hole under an overhanging oak. Beef cattle grazed around the banks. He had tried to stock the land with pheasants. "I raise pheasants and let them go. I never see them again. I don't know where they're going, probably up north looking for prairie." About half of his land was cornfields, but Hawley didn't want to farm it, so he rented the land to farmers and worked at the steel mill.

Hawley wanted to show me where the steel mill got its water—the man-ufacture of steel requires vast amounts of water. We drove through Ladoga, a little town of Victorian houses with wide verandas screened by lilacs and hydrangeas. Hawley pointed to a lovely old house, on Ladoga's one main street, surrounded by an iron picket fence. "That was my grandfather's house. He was the town doctor. My parents are living there now." We con-tinued south from Ladoga until we came to Cornstalk Creek, a honey-colored stream running over sand. Chief Peter Cornstalk had lived on the banks of the creek in a village with his people, the Miami Indians, Hawley explained. Hawley drove the truck through the creek and stopped at a steel pipe. It was a well in a Nucor well field, and it was connected to the plant by a pipeline. We got out of the truck. Deep woods shadowed the creek—sycamores and black willows—along the banks there were buttercups and

■

mayapples in bloom, and jewelweed was heading up in masses. I put my hand in the creek. A little warm for trout.

"When we dug these wells we found rocks with gold in 'em, at a hundred feet," said Hawley. "It was a gold vein." Keith Busse had ordered his engineers to ignore the gold vein and keep drilling; he told them that Nucor was in the steel business, not the gold-mining business.

We got back into the truck and Hawley drove it over a wooden covered bridge, through dappled woods, through a town called Raccoon, and past a sea of abandoned greenhouses. Young trees were shafting up through the greenhouses. "They say that Raccoon used to be the geranium capital of the world," he said. "I don't know what happened. Maybe the bottom fell out of the geranium market." We headed north by east toward the Crawfordsville Project, and pretty soon we saw it, three gray shadows visible for miles, three erections of cold metal on the bosom of Indiana, an American pastoral.

■ ■ ■

Hawley tramped down a corridor in a construction trailer, carrying a pile of paper and smoking a Macanudo cigar, knocking clots of mud from his boots. He entered an office, where he found certain Run-a-Muckers hanging out, drinking Pepsi from cans.

Hawley waved his pile of paper. "Look at these bids," he said. "These contractors are screwing me! Their bids are double my budget!"

"Try J & D," said a foreman named Kevin Perala. Perala had been a foreman at Inland Steel, on the Lake Shore, when Nucor's headhunters found him. Perala was in his late twenties. He had a mop of blond hair and close-set eyes.

"J & D, yeah, I'll try J & D," said Hawley. He accepted a Pepsi from someone. He wrapped his cigar hand around the soda can so that he gripped both the can and cigar in the same hand, cracked the can with his other hand and took a sip, and then tilted his can hand to maneuver a glistening fishtail of cigar to his lips, for a mouthful of restorative smoke. "I quit cigarettes. Now I'm inhaling Macanudos."

The walkie-talkie on Hawley's belt crackled. "Will Hawley! Will Hawley!" it cried.

He took the butt out of his mouth and unclipped a microphone from his belt. "Will, here," he said.

"Crrk, blah, blah, blah, crrk!" said the walkie-talkie.

Hawley replied, "Yeah, they're pouring it right now."

Kevin Perala turned to another foreman named Tom Weiler. "Tom, you've let million-dollar contracts, huh?"

"I've got seven contracts to deal with right now," said Tom Weiler, leaning against a Formica table, sipping a Pepsi in a cool sort of way.

"You ever drunk moonshine, Tom, with a vice-president?" said Perala.

"I worked at U.S. Steel and Bethlehem Steel," replied Weiler, "and I never even met the plant manager at either place, let alone a vice-president."

"Let alone a vice-president who takes your money at pool," said Perala.

"Will Hawley! Will Hawley!" It was Hawley's walkie-talkie again.

"Will, here."

"Crrk, blah. Crrk, crrk, crrk."

Hawley muttered a reply and hooked the microphone back on his belt. He remarked to the room, "I don't do anything but live and breathe steel. When we go fishin' and drinkin' now, all we talk about is steel."

" 'Will, Here' had to give up screwing," said Perala.

"Not yet, anyway," said Hawley.

"We call him, 'Will, Here,' " said Perala, "because that's what he calls himself when he answers his walkie-talkie—'Will, Here.' "

"I had to give up drinking on account of this job," said another guy. "I don't drink anymore, except alone or with people."

When they weren't pickling their minds in Nucors made of Scotch and gin, the Run-a-Muckers spent a lot of time reading construction magazines, hoping to learn valuable tips. They were trying to figure out how to build a steel mill. They made telephone calls all over the United States from their trailers, trying to locate contractors who could do a job that needed to be done immediately at the Crawfordsville Project.

Perala explained the Run-a-Muckers' technique. "Say I need some pipe," he said. "I don't know where to get pipe. Hell, I've never even done *construction* before. So I go through the construction magazines, and I see who's selling what. Then I call up some people and I say, 'Hey, this is Nucor! We need ten thousand feet of your pipe, *right now*. How much do you want for it?' Or we send them some drawings and a specification sheet, and we ask them, 'How much money is this?' And they give us bids. Everything here is done with bids. We bid everything lump sum. Everything. We don't do 'cost plus' or 'time and materials' jobs with contractors."

Contractors who didn't do a job right were made to tear out their work and start it over again. That was called a tear-out. There had been so many tear-outs already that Perala wondered if the job site would dissolve in a

■

donnybrook of fistfights between steelworkers and contractors. "All I know is, on the day they melt steel, half of Nucor's gonna land in the county jail."

"If we don't have a blueprint, we make a best-guess drawing and give it to the contractors," said Weiler. "This place is planned with cartoons."

Ken Kinsey, a steelworker, wandered into Perala's office. Kinsey was the individual whom Duane Hurler, the iron hanger, had threatened to kill that night in the Scoreboard Lounge—a large guy in his twenties, with gentle eyes behind eyeglasses.

"My first day on the job they threw me right into the fire," said Kinsey. "My boss said to me, 'I'm gonna give you the steel erection in the melt shop.' "

Building the melt shop required millions of dollars' worth of contracts, and Kinsey knew nothing about construction. He had been a factory worker at a Nucor steel fabrication plant up in Saint Joe, Indiana. Nucor was paying Kinsey $8.49 an hour at the Crawfordsville Project, plus overtime. "I didn't sleep a whole lot my first week on the job, but I didn't puke even once," said Kinsey. If Iverson's machine worked, then Kinsey's pay would hit $40,000 a year if he became a steelworker, or $70,000 if he became a foreman. Nucor had given Kinsey plenty of rope with which to hang himself.

Perala said, "If you screw something up, Keith Busse says, '*Don't do that again.*' You can make any mistake once with Busse. I poured a slab in the wrong place. Twenty tons of concrete. Then I poured two foundations in the wrong place. We just picked up the slab and the foundations with cranes, and moved 'em, and dropped 'em, and broke 'em up. At Big Steel they're saying, 'This is crazy, nobody builds a steel mill like this.' "

5

NUCLEAR REO

■ ■ ■

The history of the Nucor Corporation is in some respects the history of American business in the twentieth century. The seed of the Nucor Corporation was planted in the year 1896, when Ransom Eli Olds, a gadgeteer in Lansing, Michigan, installed a gasoline-vapor motor of his own design in a horseless carriage of his own design, and the motor propelled the carriage at nine miles an hour around Lansing, Michigan. It was one of the first automobiles in North America.

Olds had a vision of the future. As he put it to a group of engineers, "Did you ever stop to think what a grand thing it would be to dispense with the clanking of the horses' hooves on the city pavements, how much cleaner our streets would be, and that with rubber tires, a city may become a veritable beehive without the deafening noise of today, as well as the cracking of the whip?"

The only particle known to science that can move faster than the speed of light is a dollar. The dollar began to smell something good in the motor wagon, as the automobile was then called, and in 1899, a shock wave of startup capital suddenly flooded into motor wagon companies. Venture capitalists fell all over themselves to start motor wagon companies, when they hardly knew what a motor wagon was, and motor wagon companies appeared overnight, at the rate of two or three a week. In 1899, Ransom Eli Olds struck a deal with a group of venture capitalists in Lansing to start a motor wagon firm to be called the Olds Motor Works. Olds's main backer, Samuel Smith, got twenty thousand shares of stock in the initial capitali-

■

zation of the Olds Motor Works, while Olds himself got exactly ten shares of stock. Olds understood engines better than he did the notion of equity, of common stock. He did not fully understand that control of a company is accomplished through the ownership of its stock rather than through the power of ideas or personality, and Olds did not foresee that once a corporation is formed it takes on a life of its own. Olds handed away control of his company from the beginning by relinquishing control of the stock. It was a mistake built into the capital structure of the company, an original sin, and it led to Nucor.

Ransom Olds, now an employee of the Olds Motor Works, proceeded to invent the Oldsmobile. It was a cheap runabout with two seats, with a motor located behind the seats, and it was steered with a tiller. The Oldsmobile could go twenty-five miles an hour on level ground and it could even climb hills, provided that you drove it backwards (the Oldsmobile offered better traction in reverse). The machine was remarkably convenient.

The company's sales exploded, but as so often happens in startup companies, relations between the founding genius and his backers came unglued. The venture capitalists paid little attention to Olds, and he became a frustrated man.

In 1904, Olds resigned from the Olds Motor Works and raised venture capital for a new motor wagon company. On August 16, 1904, he founded the R. E. Olds Company, the ancestor of Nucor. The original investors in the Olds Motor Works looked on the development with horror, threatening to sue Olds for using the name "Olds" on his new company. They claimed that "Olds" was the property of their company. Rather than fight in the courts, Olds changed the name of his company to the Reo Motor Car Company—Reo for his initials, R. E. O.

The Reo Motor Car Company moved into the markets with a swiftness that surely must have astonished and terrified Olds's enemies. On September 7, Olds broke ground in Lansing for a new automobile factory. By the middle of October he had invented a new car and built a prototype of it. He called it the Reo. The Reo was a heavy touring car, with a steering wheel rather than a tiller, and with a motor packed under a hood, up front. By the middle of November, Olds had driven his Reo two thousand miles. On December 15, 1904, the factory blew the whistle on its boiler. By the tenth of January the factory was staffed with workers. On March 21, 1905, Ransom Olds declared the first shipment of finished Reos. Olds had gone from seed money to a new design for a car to a new factory to a shipment of cars in seven months. From then on the Reo Motor Car Company made touring cars. Si-

■

multaneously the Olds Motor Works faltered, declined, and was bought by General Motors. Subsequently the name Oldsmobile has prospered unto this day, although it now takes General Motors five years to design a new car and eight years to build a factory in which to make the car.

■ ■ ■

Ransom Eli Olds designed Reo motor cars from 1904 until the mid-twenties. The Reos were stout, expensive machines, with heavy gears, and they could run at forty-five miles an hour. Olds advertised his Reos in the pages of classy national magazines. In 1912, Ransom Olds introduced the Reo Thirty, "a roomy, powerful, stunning, car—for an even thousand dollars." In small print the ad went on to say that a windshield and a gas tank cost extra. Olds had discovered the merits of options.

In the middle of the twenties, Olds seemed to lose interest in automobiles. He resigned from the presidency of Reo (although he retained a seat on the board of directors) and began to speculate in Florida real estate, where he quickly lost his shirt, until in order to cover his losses he sold off much of his stock in the Reo Motor Car Company to the general public.

The Depression hit Reo hard. And yet Reo carried on, building beautiful luxury cars, the Reo Flying Cloud, the Reo Royale, losing money on the cars. At the age of seventy, Ransom Olds attempted to take control of the company away from its managers. He wanted Reo to drop the luxury cars and build a four-hundred-dollar car for the people, the cheapest car ever sold, a resurrection of the original Oldsmobile, updated for the Depression. But Olds had done it again; he had surrendered any real influence over Reo by having sold his stock. Reo's other directors resisted the idea of a cheap car. After mounting a proxy fight and losing it, Olds decided that he had "reached a point where I cannot sleep nights and unless I do something I may find myself a nervous wreck," as he put it. This time he withdrew permanently from the company. Had Olds got his way with the board, Reo Motor Car might have produced a rounded little car with teardrop fenders, and it might have been called the Oldswagon.

Reo Motor Car found better profits in trucks than in luxury cars. The Reo Speedwagon series of delivery trucks ("America's *toughest* truck!") did well, but not well enough. In 1938, Reo Motor Car filed for protection from its creditors under Chapter X of the Bankruptcy Act. Reo Motor Car was then reorganized as Reo Motors, Inc., while the company discontinued its passenger cars and became solely a manufacturer of trucks.

Reo Motors lost market share. It tried to diversify into lawnmowers,

■

which looked like a growth business after World War II. The Reo Royale, once a luxury car, became the Reo Royale, a luxury lawnmower. On the last day of December 1954, Reo gave up. It sold off all of its manufacturing operations, sustaining a three-million-dollar loss on the sale of the assets. The Reo truck line ended up with the Diamond T Truck Company, and Diamond T made Diamond Reo trucks until 1975, when Diamond T went bankrupt.

The story of Nucor continues with the remains of Reo Motors, Inc. After the sale of the truck-manufacturing operations, Reo Motors became a shell company, nothing but a set of bank accounts, with a net worth of $16 million and no business whatever. So Reo began to pay out the money in liquidating dividends, to put itself out of existence and to reward its long-suffering shareholders with cash on the barrelhead.

The liquidation of Reo went forward and the contents of the corporate treasury went out in the mail to Reo's stockholders. But the liquidation triggered a series of events in 1955 that brought the company back from the grave. Reo had a hidden asset on its books, which was the loss of $3 million that the company had sustained when it sold the truck business. The loss could be applied to any future profits that Reo might earn, thus reducing Reo's future income taxes. A group of dissident shareholders noticed the losses on Reo's books and wanted to get a tax advantage from the losses. In September 1955, the dissidents challenged the liquidation of Reo in a proxy fight that came to a climax in the ballroom of the Roosevelt Hotel in New York City, where Guy Lombardo and his Royal Canadians sometimes played. The dissidents won the proxy fight and took control of Reo, whereupon they forced Reo to take over a microscopic entity known as Nuclear Consultants, Inc. It was a forced takeover, an unusual move in corporate finance, wherein the dissidents forced Reo to take over Nuclear Consultants against Reo's wishes.

There followed a blizzard of proxy paper. Reo emerged from the blizzard as the Nuclear Corporation of America, under new management, and opened an office in the Empire State Building. The Nuclear Corporation of America had big plans. It wanted to be the General Motors of the atom. Nuclear's shares of stock were listed on the American Stock Exchange, creating a stir of excitement. The Nuclear Corporation of America was the first publicly traded nuclear company, and in the 1950s anything with the word "nuclear" attached to it sounded like the future. For a publicity stunt, the first trade of Nuclear stock, in October 1955, was actuated on the ticker tape by a burst of nuclear energy.

■

Licking their dry lips, investors dialed their stockbrokers to buy stock in the Nuclear Corporation of America, even though they had little idea of what the company did, since the company itself had little idea of what it did. The president of Nuclear was a certain Sam Norris, a man of vision. Unfortunately his vision exceeded even the powers of the atomic nucleus. Norris was stunned by the possibilities of the atom, and his company was a mooncalf, a Reo Motor Car mutated by radiation. Sam Norris got the company into every possible nuclear business. Nuclear built radiation sensors in New Jersey under the trade name "Nucor." Nuclear purchased a company known as Radioactive Products of Detroit—whatever those were. Nuclear announced plans "to carry out construction of a nuclear reactor at an early date"—which it never did. Nuclear bought a company known as Research Chemicals for $50,000. Research Chemicals had a mission to search for uses for "the 14 comparatively unknown elements known as rare earths," as the visionary Sam Norris put it.

Of all the varied businesses that the Nuclear Corporation of America established, the most effective nuclear industry it discovered was the issue of Nuclear stock. By and large the atomic businesses went nowhere, and profits, as they say on Wall Street, remained elusive. In fact, Reo-gone-Nuclear had not made a regular and sustainable profit since the twenties, around the time that Ransom Eli Olds became bored with automobiles and started to speculate in Florida land. Upon the departure of its founder, Reo had embarked on a meandering search for fortune—luxury cars, delivery trucks, luxury lawnmowers, rare earths, radiation sensors, radioactive products of Detroit—and had become lost in a maze of theories that abounded with hope and led nowhere.

■ ■ ■

Francis Kenneth Iverson grew up in Downers Grove, Illinois, then a small town, located thirty miles west of Chicago. In his youth he was known as Fran. He grew up in a moderately well-to-do family. Iverson's father was an electrical engineer who worked for the Western Electric Company. The Iverson family spent weekends and summers on a big farm on the Fox River, to the west of Downers Grove, owned by Fran's (Ken's) grandparents. The farm encompassed a thousand acres of rolling corn and sections of timber.

He liked to roam the farm, hang out by the river. "I did everything a kid would do. I smoked corn silk, I went skinny-dipping in the Fox River," said Iverson. He played the baritone saxophone in high school for a swing

■

band. He can't remember the name of the band. The band played gigs at high school and college dances around Downers Grove, specializing in "The Johnson Rag" and "String of Pearls." He won second place for baritone saxophone in a national music festival. He enrolled at Cornell University during World War II under the U.S. Navy's V12 program, and then entered the Navy as a lieutenant junior grade in the spring of 1945. The Navy sent him to Johnston Atoll, the smallest U.S. possession, a pimple of coral a quarter-mile across, sitting in the Pacific Ocean four hundred miles southwest of Hawaii. Iverson was in charge of the welfare and recreation of aircraft crews that stopped at Johnston Atoll. Three months later the war ended, and Iverson went on to graduate school at Purdue University, where he got a master's degree in mechanical engineering with a minor concentration in hot metal.

"I wanted to study something completely empirical, and that was metallurgy," as Iverson explained. "At that time there was no link between metallurgy and physics. A lot of things about steel were unexplained then, and they still aren't explained today. We know the crystalline transformations of steel, but we don't know *why* steel forms certain crystals; it just does. There is an art to making steel."

When he got out of graduate school, in 1947, Iverson found a job as a research physicist at the International Harvester Manufacturing Research Center in Chicago, where he operated an electron microscope, taking pictures of metals. It was at that time, when he entered business, that he decided he would rather be called Ken than Fran. In 1952, he took a job as the chief engineer at the Illium Corporation, a foundry in Freeport, Illinois, that operated in a five-car garage. One day the Illium Corporation received a contract to make some exotic alloy nozzle tubes of a type used to extrude pork sausage casings. The sausage-casing material was extruded into an acid bath that ate away your ordinary tool steels. The Illium Corporation had an acid-proof alloy known as Illium G, suitable for extruding the casings of pork sausages into an acid bath. Illium G, a tough metal, was made from nickel, chromium, copper, and iron. A casting machine was what the Illium Corporation needed to make sausage nozzles, but Illium, being a company in a garage, couldn't afford to buy any expensive casting machines. So Iverson designed and built a cheap casting machine for the purpose of making sausage nozzles.

Iverson's machine cost $6,000. Like all casting machines, it was a mechanical nightmare. It was what is known as a spin-casting machine. The core of the machine was a horizontal pipe, eight feet long, lined with

■

baked sand. The pipe was attached to a motor by a drive belt. When you started the motor, the pipe spun on its long axis like a pencil being rubbed between the palms of the hand, at wicked velocity, 1,200 revolutions per minute. There was a sort of horn which led inside the spinning sand-lined pipe. You were supposed to pour liquid Illium G alloy into the horn, and the horn would spray the liquid alloy into the spinning pipe. The alloy inside the spinning pipe was supposed to climb the inner walls of the pipe, form itself into a hollow tube, and crystallize, thus becoming a sausage nozzle. Then you were supposed to stop the machine, open up the pipe, knock out the sand, and out of the pipe would fall a tube made of Illium G—a sausage nozzle.

No reason why it shouldn't work. He switched on his machine for a dry cast. A dry cast is a test without liquid metal. The dry cast began, the pipe began to spin, and the machine, which sat on wheels, gave off an ominous shriek and took off across the floor on its rattling wheels, terrifying the foundry workers. They refused to touch Iverson's machine. They were afraid that the first time they poured hot metal into the machine's horn, the machine would explode and spray Illium G around the room, cutting people in half and burning them to death.

Iverson and Illium's president were the only employees of the company who dared to touch Iverson's machine. They made a heat of liquid Illium G in a furnace and then switched on Iverson's machine. The machine ramped up into an earsplitting whine, and the foundry workers gathered near the garage doors, getting ready to run for their lives.

"The pipe made a monstrous noise of metal spinning on metal," recalled Iverson. "The president and I talked casually over the sound of the machine, even though we couldn't hear each other's voices." They chatted with smiles on their faces, possibly dying of fright, while they poured hot metal into the horn. The metal ran through the horn and into the spinning pipe. Nothing happened; the machine seemed to work fine. They stopped the machine, opened up the pipe, and knocked out a sausage nozzle. The foundry workers were embarrassed and took over the operation of the machine. After that the Illium Corporation did a good business in your sausage nozzles.

The president of the Illium Corporation was so proud of Iverson's machine that he invited the company's board of directors to watch Iverson give a practical demonstration of his machine, and naturally that was when Iverson's machine exploded.

While the board of directors stood watching, Iverson switched on his ma-

■

chine. It ramped up to a whine, making the monstrous noise of metal spinning on metal, and the foundry workers poured Illium G into the horn. Suddenly there was a cracking sound. Then something dreadful happened—glowing liquid alloy began to stream *back out* the spinning pipe and out around the horn. In other words, the machine vomited. This is a classic form of indigestion in a casting machine. As Iverson describes it, "It threw liquid metal in a giant pinwheel. The pinwheel cut right through the roof of the garage."

The foundry workers roared "Aw, shit!" and threw away their tools and broke for the exits. Iverson broke and ran, and the board of directors ran for its life. ("Everybody runs for the door during an accident with liquid metal. That's what always happens.") The directors practically knocked each other down trying to escape from the garage, clawing over each other's backs, while Iverson's machine uttered a maniacal groan and fired a vicious stripping burst of hot metal through the roof. There was one minor injury. "This one guy—he was one of the pourers—happened to be wearing moccasins, when he should have been wearing boots," said Iverson. "He got some liquid alloy in his moccasin. He yelled and kicked his moccasin a hundred and fifty feet across the shop. That was the last time the board of directors ever visited the garage."

Not long after that, Iverson went to the board of directors with a proposal to build a new foundry. His idea was to pour capital into a state-of-the-art greenfield plant. The board did what boards do best: it decided to study the situation. "They said, 'We'll do it in five years.' So I left."

From Illium he came to the Cannon Muskegon Corporation, a producer of exotic nickel, iron, and cobalt alloys based in Muskegon, Michigan, where he worked as the sales manager and chief metallurgist from 1954 to 1960. One of Iverson's jobs at Cannon Muskegon was to cast uranium ingots for the fuel cells for the nuclear reactor inside the Polaris nuclear submarine. The ingots were made of uranium 238. Uranium was the ultimate hot metal, because it was radioactive; furthermore, it had to be cast into little crucibles lined with beryllium oxide, and beryllium is a deadly poison. All of the melting and casting had to be done inside a remote-controlled vacuum chamber. Iverson worked the controls of the chamber while he looked in through a quartz window at glowing molten uranium drooling into little poisonous crucibles. He wore a respirator and a white radiation suit and radiation boots. Uranium weighs one ton per cubic foot. A lump of uranium the size of *Webster's Unabridged Dictionary* would crush a desk. Iverson's ingots were four inches in diameter and they

■

weighed a hundred pounds. One day he had to hand-deliver an ingot to the Westinghouse Electric Corporation, the manufacturer of the Polaris reactor. The easiest way to make the delivery was to fly to the customer, but uranium was a dangerous contraband material banned from commercial airline flights. Iverson decided to fly with his uranium anyway. He put the four-inch ingot into a small piece of hand luggage, and it nearly dragged his hand to the floor. At the check-in counter, he could barely lift the handbag onto a weighing scale, and a ticket clerk asked him what was in the handbag. He told the clerk it was an iron ingot. If the clerk had known anything about iron, the clerk would have known that a piece of iron four inches across could not possibly weigh a hundred pounds, and that the ingot could only be uranium. Iverson kept the uranium under his seat during the flight.

Cannon Muskegon also dealt in iron alloys. The company had a casting machine that made steel shot, small pellets of exotic steel alloy that were used as raw material for aerospace investment castings. The machine rained molten alloy droplets into a tank of water, where the droplets froze into little pearls. The water in the tank tended to boil whenever molten steel rained into it, making a violent rumble that shook the building. One day Cannon Muskegon hired a new man to operate the shot-casting machine. The man had never been face to face with hot metal, but the company put him to work on the machine's pouring platform, located at the top of the machine.

The shot-casting machine started to rain liquid steel into the water, the water boiled, the machine began to shake, and the building began to shake. The guy jumped off the top of the machine and fell eight feet into a sandpit and ran out of the building, never to be seen around Cannon Muskegon again. Iverson began to appreciate the fact that some people cannot look hot metal in the eye.

The president and founder of the Cannon Muskegon Corporation was a certain George Cannon, Sr., a hot metal man of the old school. At the age of seventy, George Cannon still showed up every day at the foundry to chat with the hot metal men and to bring himself into the illuminating presence of hot metal. The old man couldn't keep away from hot metal. At night, after a dinner party, he'd show up at the foundry wearing a tuxedo. He'd take off his jacket and his cummerbund, remove his gold cuff links, roll up his sleeves, and climb inside a furnace and bang around in there with a bunch of furnace workers, helping them line the furnace with cement. That was called ramming the lining, because you had to ram the cement against the inner walls of the furnace. "George Cannon was a tough old guy," said

■

Iverson. "His whole life was the foundry. You know these old guys, how they'll grab you by the back of the arm? Right under the shoulder, when they want to tell you something? That's the way the old fellows make a point. As for myself, I had a tendency to lose my temper. I had been shouting at a man one day, and Cannon grabbed me under the arm."

The hand that rammed cement almost mangled Iverson's arm. "He said to me, 'Ken, you've got to learn how to control your temper. A good businessman is hard to bruise and quick to heal.' " Iverson came away from the experience with a bruise under his arm. He tried to remember the old man's lesson, not with perfect success. Iverson says that he no longer shouts at employees or officers of the Nucor Corporation. Iverson says that he *used* to shout at them. "I don't shout anymore," he said. "I get to talking loud."

While working at Cannon Muskegon, Iverson cast a number of two-inch-thick slabs of Rene 41, a nickel-cobalt-iron alloy that was rolled into the skin of the Mercury space capsule. A Mercury space capsule sits on display at the National Air and Space Museum in Washington, D.C., down the street from another Smithsonian hall where a horseless carriage from the year 1897 is on display, built by Ransom Eli Olds. The alloy skin of the Mercury space capsule is only thirteen thousandths of an inch thick, and yet it survived the fires of reentry from space. But it could not survive the pressure of millions of curious human fingers touching it, and so F. Kenneth Iverson's space alloy was covered with Plexiglas by the federal government.

■ ■ ■

The atomic thrust at the Nuclear Corporation of America, formerly the Reo Motor Car Company, had petered out. Meanwhile, two major stockholders had bought large chunks of Nuclear. They were The Martin Company (later to be called the Martin Marietta Corporation, the aerospace firm; it will be referred to as Martin Marietta here) and Bear, Stearns & Company, the New York investment banking firm. Martin Marietta had bought 22 percent of the stock of Nuclear, 1.4 million shares altogether, and Bear, Stearns owned 13 percent of the stock. Martin Marietta and Bear, Stearns, now in effective control of Nuclear, joined forces to try to revitalize the company. Toward this end, in November of 1960, they brought in a new chairman of the board, a cosmopolitan fellow with friendly, gracious manners and a worldly air about him, a naturalized American, born in England and raised in Canada, by the name of David A. Thomas.

David Thomas is the most important figure in the history of the Nucor Corporation, except for Ken Iverson. Thomas is a mysterious, hidden figure who lurks in the background of the company, an executive who shaped Nucor more profoundly than anyone except Iverson, although it could be said that the Nucor of today is as much a violent rejection of Thomas's management style as it is a monument to his vision. David Thomas was just the sort of chief executive that the Nuclear Corporation of America needed, in the opinion of the company's major investors. He was a dapper man who favored bright pocket squares, a member of the Burning Tree country club. He was an aviator and a gourmet who delighted in seafood, particularly in Restigouche salmon, very large Maine lobster, Belon oysters, and the rare Winnipeg goldeye. With barely silvered hair, laid flat back with tonic, David Thomas was undeniably handsome, and he carried loosely in his fingers a set of tortoise-shell half-glasses for reading financial statements or for peering at the latest inventions in aerospace technology. A slight pouch gathered under his chin and ran seamlessly into a neck that bulged a little over the back of a starched white collar, which suggested that he carried a burden upon his flesh of exquisite sauces.

David Thomas moved easily among celebrities and was something of a celebrity himself. Briefly he had been married to the lead vocalist with Benny Goodman's orchestra, the popular singer Martha Tilton—her biggest hits were "When the Angels Sing" and "Loch Lomond." He was a good friend of George M. Bunker, the chairman and chief executive of the Martin Marietta Corporation. When Thomas visited Manhattan he dined at "21" with George Bunker, where they talked about the grand future that awaited America in the aerospace age. (Thomas, who is alive and well today, now says that, in fact, he really preferred to eat at Christ Cella's— the lobsters there are enormous, he says—but that the Martin Marietta chairman insisted on taking him to "21.")

It was through George Bunker, who happened to be a director of the Nuclear Corporation of America, that Thomas got his job as the chairman of Nuclear, in 1960. As soon as he had been appointed chairman, Thomas took a quick look around at Nuclear's businesses, decided that they weren't going anywhere, and promptly encouraged Nuclear's president, the visionary Sam Norris, to resign. Mr. Norris soon obliged. Thomas then became president of Nuclear, as well as chairman.

A historian of the Nucor Corporation today is Samuel Siegel, the company's chief financial officer, who was an accountant at the Nuclear Corporation of America before it became the Nucor Corporation. Siegel is in his late fifties and possesses a flat, humorous face and a precise, deadpan manner of speech. In Sam Siegel's opinion, "David Thomas was a real promoter type. He was a good salesman, tall, charming, good looking, smart, articulate, knowledgeable, adept at human nature, amoral, and he knew from beans about operating a company.''

"Ha! Ha! Sammy Siegel!'' exclaimed Thomas, when I asked Thomas for a response to this remark. "I just don't understand the venom, here!''

David Thomas loved airplanes and he owned a twin-engined Beechcraft. It gratified him to drop out of clouds into cities that gleamed below him like diamonds, where he could stay in fine, tall hotels. "David Thomas believed in living right. On the company's money,'' claimed Sam Siegel.

Nucor's corporate memory has not been kind to David Thomas. He comes across in the Nucor recollection as a magisterial soul, with a keen sense of self-interest and an uncanny ability to get what he wanted from a board of directors. He leased his Beechcraft back to the Nuclear Corporation, collecting money from the corporation to pay the operating expenses on his airplane, with the full approval of Nuclear's board of directors. Thomas liked to fly the directors around in his plane. He had been flying since the age of sixteen, but according to one recollection, he was a bad pilot, or perhaps he just had a majestic, superb way of cutting the clouds that frightened those who did not know the air as Thomas did. At any rate Thomas's flying style made at least one of the company's directors uneasy. He warned the board that Thomas was likely to crash and burn, along with half the board of directors. That would have retarded the company's evolution, so the directors hired a pilot for Thomas, to fly him around the United States in the Beechcraft. If it occurred to the board that it would have been cheaper and perhaps faster for Thomas to fly in commercial airliners, the board was not moved to speak.

Ideas on Wall Street rotate with the velocity of beautiful women in strange clothes on the runway of a fashion show. If an ideal of classic beauty during the fifties had been the atomic nucleus, then an ideal of classic beauty during the sixties was the conglomerate. David Thomas, having decided that the atomic nucleus wasn't going anywhere, transmuted the Nuclear Corporation of America into something of a small conglomerate.

A conglomerate as it was imagined in the sixties was a kind of sheath dress in which assorted sexy businesses bulged and moved around suggestively. The conglomerate's profits were supposed to expand faster than its growth in revenues when the conglomerated businesses would begin to rub against each other through a type of mysterious friction known as "synergy.'' Given enough synergy, the conglomerate would come to a financial climax, an orgasm of cash flow, and everyone would get rich on the stock.

■

Thomas enlarged the company's revenues rapidly, from $2 million a year to $20 million a year, between 1960 and 1965, mainly by buying companies. Nuclear expanded swiftly by one full order of magnitude, like a mushroom cloud, although the expected profits did not immediately fall to the bottom line.

■ ■ ■

David Thomas snapped up all kinds of companies in friendly transactions, anything that looked promising, anything for the future. First, though, he got rid of Isotope Specialties, which wasn't making any money. Next he purchased U.S. Semiconductor Products, Inc., located in Phoenix, Arizona—this was at a time when semiconductor companies were coming of age. He then moved Nuclear's headquarters to Phoenix. However, he did not completely get out of the nuclear businesses; he held on to the nuclear instruments division, located in New Jersey, which sold the "Nucor" line of radiation sensors.

In 1962, in the small town of Florence, South Carolina, a business came up for sale that looked mighty attractive to David Thomas. It was the Vulcraft Corporation, a manufacturer of roof joists made of bar steel. Vulcraft had been founded by a man named Sanborn Chase (not related to Chase & Sanborn coffee), and Chase had died at a young age of a heart attack, leaving the company to his widow.

Now Ken Iverson enters the story. Iverson had become acquainted with Thomas through some business dealings that involved Iverson's employer, Coast Metals of Little Ferry, New Jersey. According to Iverson, Iverson checked out Vulcraft and told Thomas that it looked like a good company for Nuclear to buy.

Thomas then skillfully negotiated a deal with the widow, persuading her to sell Vulcraft to Nuclear for $1 million, a pretty good price for Nuclear. Thomas then hired Iverson to run Vulcraft as a general manager; and so in 1962 Iverson came to work for the Nuclear Corporation of America.

David Thomas thought the steel joist business looked terrific, so he built a second Vulcraft joist plant, in the small town of Norfolk, Nebraska. Thomas, still bent on growing Nuclear to greatness, bought Valley Sheet Metal, a company in Phoenix that fabricated air-conditioning ducts. Valley Sheet Metal was the largest air-conditioning duct company in Arizona, and as long as Phoenix grew, so might air-conditioning ducts. Thomas later appointed Ken Iverson to be a group vice-president in charge of both

■

Vulcraft and Valley Sheet Metal and transferred Iverson to Nuclear's Phoenix headquarters. Now Iverson was no longer a general manager. For the first time in his career Iverson had become a corporate suit, running two separate divisions from a corporate office.

The future looked fantastic to David Thomas. As Thomas expressed it in Nuclear's 1961 annual report, "Your management recognizes that we are on the frontier of new and great developments in the aerospace age and of all the many facets springing from the same."

"I would call it synergistic bullshit," said Iverson.

"There we go again, that venom!" cried Thomas. "How can anyone say a word like that, given where this nation was going in 1961? We walked on the moon after that!"

Seeking to bring Nuclear into the aerospace age, David Thomas hired a retired British airline executive and put him in a rented flat in London. "His job," said Thomas, "was to see if he could buy cheap European steel for Vulcraft. He didn't buy any, but he was working on that." Thomas called the British operation the Nuclear Corporation of America (Great Britain), Ltd. Thomas undertook development of a top-secret experimental machine for manufacturing tin cans. ("We wanted to manufacture cans with an electric seaming process," said Thomas. "We worked hard at it but we were just never able to do it.") Thomas started development of a new type of paper-copying machine, ultra–top secret. ("At that time copier machines were the hottest business you can imagine. It was well worth a try. And we came so *damned* close to success.")

"Nuclear was basically a schlock company," claimed Sam Siegel. The word *schlock* means "broken merchandise." Siegel had come to work for Nuclear as an accountant in 1961. "I really found ways of pumping up the profits," Siegel said, not without a touch of pride.

When Thomas's pilot flew him from Phoenix to New York City the flight droned on for two days. ("I flew with David Thomas to New York once," said Sam Siegel. "The word to describe that trip was *schlep*. Can you imagine schlepping in a piston-engine airplane from Phoenix to New York? We landed at night and slept somewhere. It was probably in a hotel, but I don't remember, probably because I don't want to remember.") After two aching days in the air, with touchdowns for the occasional pee, they gloriously banked around Manhattan and entered the approach pattern for LaGuardia Airport, and then the pilot, on final approach, abruptly nosed the small plane into a heart-stopping dive toward the runway that almost blew Siegel's eardrums. ("I swore I'd never fly with David Thomas again.'') (''Ha! Ha! Ha! Sammy Siegel knows absolutely *nothing* about flying.'')

Thomas booked a room in the Canadian Club, near the top of the Waldorf-Astoria Hotel in Manhattan, and dined at ''21'' with George Bunker, the chairman of Martin Marietta. Sam Siegel crept into a sublunary room at the Waldorf and avoided ''21'' like the plague.

There was an employee at the Vulcraft joist division, a big, quiet South Carolinian, a working man from a dirt-poor background, who had started out as a welder at Vulcraft, welding chords and trusses of bar steel together to make roof joists for shopping malls and gymnasiums. His name was Hugh David Aycock. Hugh David Aycock is now the president and chief operating officer of the Nucor Corporation. Aycock is known to everyone in the company as Dave. Thomas was much impressed with Dave Aycock: ''One hell of a smart fellow; I liked him tremendously.''

Aycock showed great skill at managing factory operations. The workers respected Aycock and would do anything for Aycock. Thomas promoted Aycock rapidly, and by 1962 Aycock had become the plant manager of the new Vulcraft plant in Norfolk, Nebraska. And so by the early sixties, the current management team of the Nucor Corporation—Ken Iverson, Sam Siegel, and Dave Aycock—all had a place inside the company, and they all happened to know one another.

Profits remained elusive. The stock price settled down to the equivalent of one dollar a share, unresurrectable by any more bursts of nuclear energy or financial synergy. Meanwhile the company was bleeding cash extravagantly. The real catastrophe turned out to be the Valley Sheet Metal division, the air-conditioning duct company in Phoenix. It went into the red soon after Thomas bought it, and then the losses at Valley Sheet Metal began to accelerate, cash flow running fiercely negative, until it looked like the Nuclear Corporation of America would be sucked into an air-conditioning duct and lost.

''It was a bad decision to buy Valley Sheet Metal,'' admitted David Thomas. ''I take full responsibility for that decision. But who do you think was the manager of that division? *Iverson.* That was the source of the trouble. There wasn't any trouble otherwise. We were strangling on the losses in Iverson's division. That division went to hell under Iverson, and the records show it. Iverson took contracts for air-conditioning projects that just drove us crazy. He lost *so* much money on those contracts. He had done well with the Vulcraft business, but I have to tell

■

you that Vulcraft had a hell of a good staff. But Valley Sheet Metal was another story, it had a weak staff with Iverson running it. He politicked. He disliked supervision. He was very much against supervision. I really had to get tough with him. He was hard-headed and he was hot-headed. Iverson has got quite a bad temper. He clashed with everyone in the company. He clashed with people until the room shook."

On January 1, 1965, the company's evidently satisfied board of directors gave David Thomas a large raise and a ten-year employment contract. The contract contained a golden parachute that would award Thomas cash payments of $250,000 in the event that Thomas ever happened to be fired. Three months later, in March 1965, Nuclear defaulted on two bank loans and headed straight for bankruptcy.

■ ■ ■

It is not in the nature of boards to know what is going on. Nuclear's board of directors was small and was dominated by its chairman, David Thomas, a stylish and confident man. Before a company's problems become obvious to the world and suspected by the board of directors, the first sign of trouble is heavy occult bleeding of résumés. Ken Iverson and Sam Siegel both independently decided to quit Nuclear but didn't tell each other or anyone else. A couple of months after the board gave the chairman his ten-year contract, Iverson and Siegel bumped into each other in a post office in Phoenix, where both of them were putting huge stacks of their résumés into the mail. They burst out laughing.

Nuclear's big investors didn't think it was so funny. When Nuclear defaulted on its debt, it suddenly dawned on George M. Bunker, the chairman of Martin Marietta, the controlling shareholder, that David Thomas's company was in trouble. This was a delicate situation, for Bunker regarded Thomas as a friend. But Bunker decided that Martin Marietta had better cut its connections with Nuclear quickly. So Martin Marietta, having bought 1.4 million shares of Nuclear at a high price, became extremely itchy to sell 1.4 million shares of Nuclear at any price at all. Buy high, sell low. George Bunker didn't care what kind of money he got for Martin Marietta's twenty-two percent block of Nuclear, he just wanted to dump the stock off his company's books immediately, before Nuclear went bankrupt, so that Martin Marietta would not be tainted by a bankruptcy.

David Thomas pleaded with George Bunker not to sell off the Nuclear shares. He predicted that the Nuclear shares would be of great value to Martin Marietta someday. He begged Bunker for just a little more time, a little more cash. He told Bunker that if he could only have a million dollars in extra capital, he could pull Nuclear out of its troubles. Bunker refused to put up a single extra dollar for Nuclear. Bunker was determined to cut his losses.

In cases like this, there is always a ragpicker waiting in the shadows.

David Thomas saw the ragpicker make his move. Thomas had dropped by Bunker's office in Manhattan, at Martin Marietta's headquarters, in the late spring of 1965, after Nuclear had defaulted on its debt, to have a cocktail with Bunker in his office before they both went to lunch at "21." "George Bunker generally had a drink in his office before lunch," Thomas recalled. "I don't know whether he kept a bottle in his desk or what. He was not an alcoholic, mind you—he was a sociable club man. He generally had a couple before lunch and a couple before dinner." The telephone rang.

The caller was a certain Donald C. Lillis, who happened to be a director of Nuclear as well as a limited partner at Bear, Stearns & Co. Lillis was calling from his office near Wall Street. Lillis had sat through Nuclear's board meetings, watching the meltdown occur, and now he had smelled an opportunity. "Anyway, the telephone rang," Thomas recalled, "and it was Donald Lillis on the line. He says, 'George, if you're going to get rid of your shares in Nuclear, I'll buy them.' And Bunker says, 'You can *have* 'em!' "

Nuclear's common stock was then trading on the American Stock Exchange at $1.60 a share. Lillis offered to pay a nickel a share and Bunker promptly accepted the offer. As Thomas recalls it, "Lillis was as cold-nosed as anybody you met in your life. And Lillis said, 'I want the transaction done in New Jersey.' "

There were two reasons for doing it in New Jersey. The first was that it would save Lillis money on taxes. The second was that it would conceal the transaction from the public markets. If word had gone around that 22 percent of Nuclear had just traded hands at a nickel a share, it might have triggered a Nuclear panic on the American Stock Exchange.

Since Nuclear's stock was then trading openly at $1.60, and since Donald Lillis had offered a nickel, you could say that the spread between the ask-price and the bid-price was 97 percent. Ask a $1.60, bid a nickel. George Bunker chased the downtick that morning and hit the bid, unloading 1.4 million shares of Nuclear on Donald Lillis for 5 cents each. Lillis snapped up Martin Marietta's whole wad for $70,000, a preprandial flea market close-out. Those 1.4 million Nuclear shares were supposedly worth

■

$2.2 million, not $70,000, if you believed that the public trading of the stock reflected any kind of reality. Donald Lillis already owned a bit of Nuclear—about 2 percent of the stock—and together with Martin Marietta's fire-sale block, Lillis wound up owning 24 percent of the company. What he got was effective control of the Nuclear Corporation of America, for whatever that was worth.

Although Lillis's purchase of the Nuclear stock was a private transaction, done quietly in New Jersey, it was in fact perfectly legal. It merely happened that two of the more sophisticated owners of Nuclear shares, George Bunker and Donald Lillis, set a private market value on the stock at a nickel when the public thought it was worth more than a dollar. A slug of merchandise had come to market, it was rotting and had to be sold. Lillis probably thought he'd got a good deal—he probably thought his Nuclear shares were worth at least twice what he had paid for them.

When Martin Marietta bailed out, David Thomas's job with Nuclear became doomed. So now the board of Nuclear had to discuss with its chairman the subject of his own firing, and unluckily the board had just given Thomas that ten-year employment contract with a golden parachute, and the contract had nine years and nine months left to run. Of course Thomas was reluctant to fire himself without the required payment of $250,000, but there was a slight problem in that Nuclear had run out of money. Thomas agreed to accept cash payments totaling $180,269—he settled for less than the full value of his contract—and his severance payments included a payment from Nuclear to Thomas to break Nuclear's lease on Thomas's airplane, since, understandably, the chairman wanted his Beechcraft back from the company. On May 27, 1965, David Thomas resigned from the chairmanship and presidency of the Nuclear Corporation of America. The settlement left Nuclear without an airplane, without cash, without a president, without hope. Another two months went by, during which the board of directors searched in vain for a new president. No one wanted the job. Sam Siegel quit his accounting job at Nuclear (no, he didn't want to be president), but on July 30 he sent a telegram to the board of directors that went, "I WOULD CONSIDER CONTINUING WITH NUCLEAR IF THE FOLLOWING OCCURRED: (1) KEN IVERSON IS GIVEN AN EMPLOYMENT CONTRACT AS PRESIDENT OF NUCLEAR. (2) I AM GIVEN A POSITION AS TREASURER AND CONTROLLER OF NUCLEAR."

The board took Siegel's terms, and on August 9, 1965, Iverson was appointed president and Siegel became the chief financial officer. David Thomas left his office the day he resigned and never set foot on the

■

premises of Nuclear again. He went to Puerto Vallarta, Mexico, where he bought land. He bought a small textile factory in Guadalajara, Mexico. He bought the Oceano Hotel in Puerto Vallarta. Thomas's Oceano Hotel was one of the most famous hotels in Mexico. Richard Burton and Elizabeth Taylor stayed at the Oceano while he was shooting *The Night of the Iguana* and while their marriage was breaking up. Burton spent a lot of time at one end of the bar at the Oceano, day and night, and later Burton and Taylor met at the Oceano bar with their lawyers to hash out their divorce. George C. Scott could be found sitting at one end of the Oceano bar and John Huston at the other end of the bar, but the two men never talked with each other, never looked at each other. At the Oceano, at various times, you could also find John Wayne. Richard Nixon delivered a speech from the steps of the Oceano. James Earl Ray stayed there in the spring of 1968, just before he headed north to Memphis to assassinate Martin Luther King, Jr.

David Thomas found his own genius in running a Mexican hotel. By his own account he has become a wealthy man. He got the best kind of revenge against Iverson and Siegel by living well ever afterward. ("Very well, very *very* well," he said.) Thomas now divides his time among the cities of Scottsdale, Puerto Vallarta, and Victoria, British Columbia. The thought of what might have been sometimes tantalizes Thomas, for if George Bunker had not become disillusioned and sold off Martin Marietta's Nuclear shares for $70,000, then Martin Marietta, an aerospace company, might very well control the Nucor Corporation today. Electric steel might have been one of the new and great developments in the aerospace age, in the words of David A. Thomas.

Thomas himself is quite well aware of the irony of that lost future. In later years, after Nucor had arisen from the ashes of Nuclear, George Bunker—still the chairman of Martin Marietta—occasionally used to visit Thomas at a massive villa that Thomas owned in Puerto Vallarta, overlooking the Bay of Flags, shaded by green foliage. As Thomas recalls it, he and Bunker were sitting in chairs reading books one afternoon, watching the sun go down over the Pacific. The surf was roaring. Thomas got up to mix himself a drink. He turned to Bunker and said, "George, you made a *horse's ass* move on that one, selling those shares to Lillis. It's your own damned fault! All you had to do was give me that million dollars I wanted, and I could have gone on. Then you'd see exactly the same Nucor that you see today, except that *you* would own it! Think of the value, George, of that Nucor stock today!"

■

Bunker put down his book. "Boy, you are so right," Bunker said. Bunker got up and poured himself a Scotch.

Thomas has always regarded Iverson and Siegel's ascendency at Nuclear as a palace coup, and he believes that he himself would have turned Nuclear into the hottest steel company in America if not for F. Kenneth Iverson. And one thing is for sure, Iverson would have been fired.

"It was *my* idea that Nuclear should get into the steelmaking business," he said. "I already had this steelmaking thing well under way, when I left. Iverson fought the idea that we should get into the steelmaking business. I made trips to Quebec and to Vancouver, looking for steel mills or raw materials to buy. Iverson argued like hell with me over that idea. I finally had to *order* him to go with me on these trips. We never did buy a steel mill, but we were heading in that direction. But the moment I left, he pursued it like hell. I once asked Iverson why he stabbed me in the back like that. Iverson was red-faced. He replied that I was a father image to him. I never understood that response from Iverson, that I was a father image to him. But I will say this for Iverson, he has done one hell of a good job on following through with the program that I established. Iverson just finished off what I started. And the rest is history."

■ ■ ■

■

6

BUILDING
THE
CULTURE

∎ ∎ ∎

I became president by default," was how Iverson put it. "No one else wanted the job." His job description was merely to stave off bankruptcy. Siegel believes that on the day that Iverson took control, Nuclear was ready for a filing for protection from its creditors under Chapter 11 of the bankruptcy code. Iverson and Siegel scrambled to get a commercial loan from a finance company. They figured that they could use the finance company's loan to pay off the defaulted bank loans. It was like trying to pay off your Visa with your MasterCard, but they managed to do just that. They got a loan from a commercial finance company on August 13, four days after Iverson became president, by pledging virtually all of Nuclear's assets as collateral for the loan. If they hadn't found the money quickly, it would have been the end of a long and twisted road that had begun with the Reo Motor Car Company. The finance company, as Siegel puts it, "took practically everything but the shoes off our feet. We really mortgaged the company."

Iverson and Siegel next took a look around to see if there was anything they could sell to raise a little cash. Available for immediate sale were the many facets springing from the aerospace age. They sold off the division known as U.S. Semiconductor. They put Valley Sheet Metal into straight liquidation. They canceled the secret research into the tin can manufacturing machine. They shut down the secret paper-copying machine. They fired the chap in London and thereby liquidated the Nuclear Corporation of America (Great Britain), Ltd. These actions stopped the bleeding and

∎

raised a pittance of cash. During the liquidations, Nuclear's book value crashed from 42 cents a share to 11 cents a share.

Now what? With a book value of 11 cents a share, Nuclear formerly Reo did not exactly have working capital. The company's only profitable unit was the Vulcraft joist division, with factories in South Carolina and Nebraska, a business that David Thomas had invested in with shrewdness and foresight. The roofs of shopping malls had nothing to do with nuclear energy or the aerospace age, but at least they earned money. Iverson felt that the Nuclear Corporation of America had to concentrate on steel joists, and so he elected to move the company's headquarters from Phoenix to Charlotte, North Carolina, to be closer to the Vulcraft plant in South Carolina.

Iverson and Siegel made an exploratory trip in January of 1966 to Charlotte, where they discovered and inspected the Cotswold Building, and rented an office on the top floor. There they installed a card table, two chairs, and a telephone, and they declared that to be the national headquarters of the Nuclear Corporation of America. Then they returned to Phoenix to move the corporate staff to Charlotte.

The corporate staff refused to go to Charlotte. Nobody wanted to go to North Carolina. "The staff thought the company would go broke," says Iverson. And so the entire staff quit. They handed in their resignations and went home.

After the departure of Nuclear's corporate staff, Iverson and Siegel were left alone in the Phoenix offices, the last two executives of the Nuclear Corporation of America. They faced row upon row of old filing cabinets, the corporate records of the Reo Motor Car Company and its successors. In the background, in the empty offices, the telephones rang continually—creditors wanting their money. Iverson answered the phones and talked fast to creditors while moving men came through the offices and carried out the Reo filing cabinets, as well as a number of old Steelcase desks (they were gray, with rounded corners, made of steel, the kind of desk you would find in a gym coach's office), and the moving men loaded the filing cabinets and the desks into two moving vans.

There was a stockholders' meeting coming up in New York City and Iverson was going to have a lot of explaining to do, because at the last stockholders' meeting, the board of directors had given the stockholders every reason to think that the future had never looked brighter for the Nuclear Corporation of America. Leaving their families behind in Phoenix, Iverson and Siegel flew to New York City for the annual meeting, in the spring of 1966, to explain to the stockholders what had happened.

The stockholders were enraged, and raked Iverson over the coals. They shouted at Iverson, demanding to know why the Nuclear Corporation of America had suddenly almost ceased to exist, when everything had seemed so lovely last year.

"It's a rotten company, what can I say?" Iverson replied to them.

He and Sam Siegel continued on to Charlotte, where they rented an apartment together to conserve personal cash, but to conserve executive strength Siegel bought expensive cuts of beef and lamb for himself and the president, shaking powdered garlic all over the meat until it was as white as a snowdrift, and then sliding the meat under the broiler on high. The president of the Nuclear Corporation of America began to complain about excessive garlic powder in the chief financial officer's cooking, but Siegel wouldn't lighten up on the powder. ("He really flooded the steaks with garlic," says Iverson. "I haven't been able to stand garlic to this day.") The Reo files and the steel gym-coach desks arrived at the Cotswold Building in the moving vans, and the gym-coach desks are still in use everywhere at Nucor's headquarters. At a cost of three dollars, Iverson invested in an umbrella tree for the lobby, with shiny leaves that were nearly as durable as plastic, and he hired a receptionist by the name of Betsy Liberman. Betsy Liberman polished the leaves of the umbrella tree to help it to grow. Betsy Liberman is still the receptionist in Nucor's offices at the Cotswold Building and she still polishes the umbrella tree, but it hasn't grown much, as of this writing.

■ ■ ■

The Vulcraft joist business provided positive cash flow, which turned into reportable earnings for the Nuclear Corporation of America. Nuclear earned its profit by purchasing bar steel from large American steel companies, welding the bar steel into roof joists, and selling the joists to builders of shopping malls and high-school gymnasiums under the Vulcraft brand name.

Nuclear's main supplier of bar steel was United States Steel. Much to the concern of Ken Iverson, U.S. Steel kept raising the price of its bar steel, until U.S. Steel was charging more for a ton of bar steel than Nuclear could get for a ton of finished roof joists. His company now threatened with extinction because of the high price of American steel, Iverson began

■

to travel in Europe, trolling for cheap steel, and he bought steel in Germany, Luxembourg, Belgium, and Poland. The Polish steel was accidentally mislabeled by the Polish supplier, and when the steel went into the roofs of shopping centers it buckled and the shopping centers drooped. The Nuclear Corporation of America had to send out a fleet of trucks to take back the Polish steel from customers.

"We didn't tell them it was Polish steel," said Iverson.

It was not a good situation, so Iverson went back to U.S. Steel as a customer. This time he bargained hard and got a contract in which U.S. Steel agreed to supply Nuclear with bar steel at a price that was slightly lower than the dockside price of the cheapest imported steel. It was a sweet deal for the Nuclear Corporation of America.

Even so, Iverson didn't like having to buy steel from anybody; he thought it made financial sense for Nuclear to make its own steel, and so he decided to build a small steel mill to supply bar steel to the Vulcraft plants. Bar steel, known also as your long bar or your merchant bar, consists of round bars, flat bars, channels, and angle irons—skinny pieces of steel, usually twenty or forty feet long. Iverson hoped to sell the mill's excess production of bar steel on the American market. Iverson now claims that, in the beginning, he had no idea that Nuclear was going to one day hang up and gut Big Steel's markets for bar steel. "We wanted to put something together to compete with the foreigners," he said. "They were the ones we really targeted."

He could not afford to build a blast furnace for smelting iron ore—a blast furnace can easily cost $200 million—but there was an alternative, and that was an electric arc furnace, a kettle for melting metal scrap. It was much cheaper to make steel out of melted junk than out of iron ore. Iverson decided to locate the miniature steel mill in the town of Darlington, South Carolina, in the middle of soybean fields, not far from the original Vulcraft joist plant. Iverson borrowed $6 million from the Wachovia Bank, in Charlotte, and began construction of the steel mill in 1968. Nuclear at that time was such a small company that a default on a six-million-dollar loan could have driven the company into bankruptcy. Nuclear was far too small a company to be building steel mills, and Iverson knew it at the time. "We played 'Bet-the-Company,' " as Iverson put it. It was not entirely a cool or rational bet, because Iverson merely wanted to save a little money on his raw material costs, even though he already had a cheap and evidently assured supply of American steel. What Iverson really wanted to do was to make steel, because he was a hot metal man.

■

He also wanted to leapfrog the American steel industry. So he equipped the mill with what were then the newest machines available—a Whiting electric arc furnace, a continuous-casting machine built by Concast, Inc., and a Swedish rolling mill that was the first of its kind in the western hemisphere. The Swedish rolling mill could be operated by just one steelworker. The Concast casting machine was a sweet little mechanical nightmare only four stories tall, and it was supposed to extrude a billet, an endless strand of steel as thick as a railroad tie, which could then be cut up into lengths and rolled through the Swedish rolling mill into bars for roof joists.

The Nuclear Corporation of America melted its first heat of steel on June 26, 1969, at the new minimill in Darlington, South Carolina. "We had about twenty steelworkers then," says Iverson. "They were carpenters and butchers and sharecroppers. We had one guy who was an adding machine salesman." It was a nonunion work force, utterly innocent of contact with liquid steel. On the day of the first heat, the sharecroppers moved nervously around the melt shop, smoking cigarettes and frequently visiting the lavatory. A crane dumped a mountain of shredded junk into the arc furnace and Iverson gave an order for the startup to commence.

The arc furnace was shaped like a pot, and it had three carbon electrodes that could be lowered into the pot to cook the junk inside it. The electrodes descended into the junk with a great fizz, there was a flash brighter than the sun, and a mushroom cloud punched to the roof of the steel mill. It was what they call a long-arc meltdown. Iverson had tapped twelve million watts of energy out of Duke Power at the flip of a switch. The drain sent echoes back through the power grid, causing lights to flicker all over South Carolina.

Long blue arcs snaked through the mountain of busted cars and smashed industrial machinery. The roar was loud enough to cause a breathing arrest in an infirm person. The steelworkers couldn't hear their own voices screaming in terror over the noise of melting steel. The noise seemed to open the sutures in their skulls, and a musky odor filled the building, the reek of a long-arc meltdown.

They began to run from one side of the melt shop to the other, yelling confused orders at each other. Iverson stood with his hands in his pockets, enjoying the show. When the junk had merged into a bath, the arc furnace was tipped over and liquid steel ran into a ladle, like tea running from a teapot into a cup. That is known as tapping the heat. Then, in a normal occurrence at the tapping of the heat, a blossom of sparks exploded from

■

the ladle and rained slowly down through the building, drenching the sharecroppers in fire.

They really started screaming now. Suddenly they quit. The entire furnace crew threw away its tools and ran into the soybean fields. Once they had got into the fields they gathered in a knot, squinting at the melt shop, expecting it to blow up. Their words were not recorded, but they probably were something to the effect that that Iverson guy in there is a lunatic, and that ain't no steel mill, either, it's a God damned nuclear accident.

Iverson was left alone beside the furnace with one foreman. Iverson didn't know what to do. His work force had just quit, and there was a ladle of hot metal sitting on the floor. The foreman had an idea. The foreman ordered the ladle to be poured back into the furnace, and then, once the furnace was engorged with hot metal, the foreman climbed onto the lip of the furnace and walked in a circle around the lip, so that the sharecroppers could see him outlined against the glare. "When they saw him do that, they kind of crept back into the building," said Iverson. "My foreman poured the steel back into the ladle again." That is, he tapped the heat again, and once again the sharecroppers were drenched in sparks, but this time nobody ran out of the building. "A crane took the ladle up to the casting tower, and we poured steel into the casting machine."

At the top of the casting machine there was a funnel for gathering the flow of liquid steel. The funnel captured the flow as it drained from the ladle. Then from the base of the funnel a strand of solidifying steel began to emerge. Suddenly the casting machine melted down. It drenched itself with liquid steel. That is, it broke out. A hot metal cascade splashed down three stories through the casting tower with a great uproar and a cloud of smoke. The sharecroppers ran back and forth, gasping with fear, lost in smoke, while foremen shouted, "Breakout! Breakout!" and sirens went off, *whoop, whoop, whoop!*

The emergency cry "Breakout!" means that steel has got loose inside a casting machine. Breakouts are not uncommon in steel-casting machines. A typical breakout happens when a strand of partly solidified steel, inside the machine, breaks apart. The strand has a liquid core and a thin skin, like a water balloon tensely full of water. The balloon pops and liquid steel gushes down through the machinery with a deafening roar and a great billowing rush of smoke, steam, and flames. The fires die down and the machine cools off, exhaling clouds of smoke. It has now had its guts welded together with steel. Steelworkers climb inside the machine and go to work with wrecking bars, cold chisels, and cutting torches, prying and

■

chipping and slicing gobbets of steel away from the casting machine's rollers.

It took the farmers and the adding machine salesman about a year and a half to learn how to handle hot metal. The plant startup dragged on for month after month, putting the Nuclear Corporation of America under increasing financial strain. "We had immense problems with the casting machine," said Iverson. "We wondered if it would ever work. I was convinced it would work, but I was down there on the floor of the mill watching the machine go through breakout after breakout."

Simultaneously the Nuclear Corporation of America was having a breakout of cash. The company blew cash like an open fire hydrant, to the point where a default on the debt followed by a quick filing for bankruptcy was beginning to look like a real possibility to Sam Siegel. "It could have resulted in Chapter 11," says Siegel. "To a fair extent the success of the steel mill became a question of the existence of Nuclear. We were losing money like it was going out of style. Ken and I were at it pretty good. Ken is subject to shouting. In that respect he is very much like my wife—bright, articulate, and subject to shouting."

United States Steel, which had been selling cheap bar steel to Nuclear, canceled the contract when it learned that Nuclear had built a steel mill. Nuclear's executives began to wonder if Iverson had been an idiot to think that Nuclear could take on Big Steel as well as foreign steelmakers with the help of a six-million-dollar loan from a local Carolina bank. The executives got into loud arguments in the offices of the rented suite in the Cotswold Building, some claiming that Nuclear was going to be destroyed by this psychotic venture into the primary steelmaking business, others claiming that Nuclear had no choice but to make steel. Iverson got to talking loud with his officers.

"We damn near got into fistfights," recalled David Aycock, the ex-welder and Nucor's president. "There were times when only a table stopped us from takin' swings at each other. We were building our culture, then, only we didn't know it at the time."

"People in the office thought Ken and I were going to kill each other," recalled Sam Siegel. "I am not big on relieving a situation with my fists, and so there were times when I decided that I had to leave a room."

Nuclear's stock, having risen to over $3 a share, cratered to 97 cents, and Siegel sweated every time he had to write a check to cover an interest payment on the debt. He feared that the check would bounce. Iverson had bet the company on a steel mill and now it looked like Iverson might lose his bet on a bounced interest check and a heap of boogered-up metal crapped out of a four-story mechanical nightmare. The casting machine just wouldn't work, kept breaking out, pouring liquid steel all over itself and down through the casting tower, with farmers shouting "Breakout! Breakout! The sucker broke out!" while steam, smoke, and flames rolled around the farmers, mixed with the *whoop, whoop, whoop!* of sirens and an eerie noise, the sizzling, popping, boiling-fat cough of hot metal burning up a machine. It was a steel mill in deep trouble.

Iverson began to be troubled by insomnia. He felt drawn to the steel mill and its problems. In the fall and winter of 1969, he kept driving down to Darlington from Charlotte, staying at the mill all night; since he wasn't sleeping anyway, he might as well be down at the mill.

"Night after night," Iverson recalled, "I'd be down on the floor at midnight, while the people were trying to get the caster to work, and we had been having breakout after breakout. One day a fellow came in wearing a black leather jacket, riding a motorcycle."

The guy was a biker, looking for a job. "He had never been in a steel mill before," said Iverson. The plant manager hired the biker on the spot and put him to work on the casting machine.

"On this kid's first day at work, we got a good cast going," said Iverson. A red-hot billet of steel began to emerge slowly from the base of the machine, like a strand of spaghetti coming out of a pasta machine, and everyone got that sweet feeling that happens to hot metal men when steel comes out of a machine.

"Then some hydraulic valves blew off the machine," said Iverson. A couple of broken hoses squirmed around, lashing hydraulic fluid across red-hot steel, sending up clouds of smoke. "This kid straddled a hot billet of steel—he stood with his legs apart over a piece of red-hot steel coming out of the caster—and he held the broken hydraulic hoses together with his hands. I was terrified he would burn himself. The hoses blew apart again, and he got hydraulic fluid in his eyes. We took him down to wash his eyes out, and he turned to me and said, 'This is the best job I ever had!' He was a hot metal man, only he didn't know it."

Despite Siegel's worry that Nuclear might default on its debt, Nuclear never went into a net loss during a quarter, and toward the end of the year 1970, the steelworkers finally taught themselves how to make steel. At first the mill sold bar steel to Nuclear's Vulcraft joist division, but as production picked up, the mill began to sell bar steel up and down the eastern seaboard of the United States, and the mill became extraordinarily profitable for the

■

Nuclear Corporation of America. In 1972, Iverson changed the name of the company to Nucor, since there wasn't anything nuclear about plain carbon steel. Nucor's earnings, which had been depressed by the startup, shot up 140 percent between 1970 and 1971, from $1.1 million to $2.7 million, and then went up 70 percent the following year, to $4.6 million. The steel mill in Darlington was a gold mine.

In 1974, Iverson completed a second electric steel minimill, in Norfolk, Nebraska. In 1975, he completed a third electric steel minimill, in Jewett, Texas. In 1977, he decided that the Darlington minimill had become decrepit—it was nine years old—and he tore it apart and rebuilt it. In 1980, he built Nucor's fourth electric steel minimill, in Plymouth, Utah. Meanwhile he was also building joist factories, and as Nucor moved into the joist business, Republic Steel, Bethlehem Steel, and other Big Steel companies moved out of the joist business. Today one out of three new factories, shopping malls, and office buildings has a roof or floor supported by Nucor Vulcraft steel. Then Iverson started building decking plants. Steel decking, which is made from sheet steel that Nucor purchases from Big Steel companies, is placed on top of joists for use as a floor support in office buildings. Nucor's earnings hit big new highs in the later seventies—$8 million, $12 million, $25 million, $42 million—geometric growth in profits, and the stock price soared. Then Nucor's growth leveled off, as other minimill companies began to put pressure on Nucor's profit margins.

The American minimills destroyed Big Steel's markets for bar steel and roof joists. They also essentially killed the importation of bar steel into the United States. No foreign steel companies could compete with the small American steel mills on their own turf, because the little American mills were arguably the most efficient steel mills anywhere in the world. But the Nucor Corporation now faced vicious competition from other nonunion American minimills. The little mills haunted Nucor in the way that Nucor haunted Big Steel; and some of the little mills weren't so little, either. Nucor's competitors made bar steel under aggressive, local names—Hurricane Industries, Border Steel, Bayou Steel, Chaparral Steel, Florida Steel, Cascade Steel Company, Seattle Steel, Birmingham Steel, Roanoke Electric Steel, North Star Steel. Nucor was trapped in the borderland, between bar steel and greatness.

■ ■ ■

The American steel industry emerged from the Second World War the largest and most modern steel industry on earth, engorged with capital and boasting the latest machines. In the year 1951, Benjamin Fairless, the chairman of United States Steel, was pleased to reflect: "Americans don't like to take second place in any league, so they expect their steel industry to be bigger and more productive than the steel industry of any other nation on earth. It is; but what many Americans do not know is that their own steel industry is bigger than those of all other nations on earth put together."

Life went along beautifully for Big Steel's executives, on the golf courses in Bermuda and at the private hunting lodges in Canada, but out in the mills, something began to chatter and click—the subdued complaint of machinery growing old. Meanwhile the Europeans and the Japanese rebuilt their steel industries out of the rubble of the war, trying to leapfrog American steel with new inventions, with new ways of making steel, with oxygen furnaces, and with electric steel minimills. American steel executives weren't interested in new inventions; in their opinion the World War II blast furnaces worked fine. When they did install new equipment in a mill, often it was a mature generation of machinery, just as a new generation of machinery was coming onstream in Europe and Japan. It seemed that the managements at Big Steel had become reluctant to bet money on inventions. They stopped taking risks. They couldn't bring themselves to invest in anything they didn't recognize. In trying to avert risk, they took huge unacknowledged risks.

The chattering grew louder, there was a smell of frying brakes, a wheel hopped the track, and Big Steel toppled over like a freight train and went into a crawling, elaborate wreck that dragged on and on for thirty years, and ended with a chain of dramatic explosions. As the eighties dawned, 320,000 American steelworkers lost their jobs in a wave of bankruptcies and plant closings. Seventy-five percent of all steelworkers in America lost their jobs. What happened was simple. The steel industry became the *world* steel industry, but the Americans didn't notice.

There are altogether too many reasons for the collapse of the American steel industry—greedy labor unions, bungled government policy, inflation, foreign competition—but ultimately it was a failure of management. The managers came up through the best business schools in America and up through the soundproofed meeting rooms of a corporate bureaucracy, but the one thing they never understood was how to invest money shrewdly in manufacturing technology, and their task of managing a giant steel company was not made easier by the fact that they never much cared for the men on the bottom who made the steel. They were never able to inspire the hot metal men to do much of anything except to go out on strike.

Between 1977 and 1987, fifty million tons of annual capacity were stripped out of Big Steel. Fifty million tons. Fifty Crawfordsville Projects. It is safe to say that not one major steel company in America did not stare bankruptcy in the face. One-fourth of the industry did go bankrupt. The Americans ended up supplying only eleven percent of the world's need for steel. The great integrated steel mills that stretched for miles along the Monongahela River, and along the Ohio River, and in Cleveland, Birmingham, Buffalo, Houston, Seattle, and Saint Louis became iron tombs infested with sumac and burrs, as if the mills had been seared by neutron bombs.

Big Steel, shrunken and humbled but by no means ready to die, had survived a devastating war. Big Steel's condition had stabilized, the Lake Shore mills were making steel more efficiently than ever, but there was an undercurrent of nervousness at Big Steel. The nervousness concerned the small American steel companies, and in particular the Nucor Corporation. What if Iverson cloned his machine at scattered locations all over the United States? One or two giant American steel companies might go bankrupt or have to take whole operating divisions out behind the barn to be shot. Big Steel executives were taking precautions. They had started a whisper campaign with Wall Street financial analysts, saying that Iverson's machine would almost certainly fail. They had told potential customers of Nucor that even if Iverson's machine worked, it would manufacture brittle, speckled, warped, worthless sheets of steel—Distressed Material, DM. You want to buy DM from Nucor? Because that's probably what you're going to get from Nucor, DM. If Iverson's machine manufactured a million tons of DM, then Nucor would go into financial shock, Nucor would struggle to refinance its short-term debt, Nucor's stock would drop like a brick on the New York Stock Exchange, and Ken Iverson's foray into the sheet steel market would turn into a massive write-off for Nucor. In fact, Iverson might be forced to retire, since he was sixty-three years old. That was something they could look forward to! So long, Ken Iverson! F. Kenneth Iverson, a footnote to American steel!

■ ■ ■

Sanborn Chase, no relation to the coffee, who had been the founder of the Vulcraft joist division of Nucor, had instituted production bonuses with his workers. The Vulcraft workers worked in teams, and if the team made a lot of joists, the team got a cash bonus. Iverson decided to stick with the bonus system and the work teams.

At first the bonuses were small, but Iverson says that he was surprised to see the bonuses becoming larger every year as the teams, now gradually becoming hot metal men, discovered ways to increase the production of steel in order to raise their bonuses. Iverson says that he was shocked when steelworkers' bonuses began to exceed one hundred percent of base pay, but since the system was bringing large productivity gains, he decided not to touch it.

The Nucor steelworkers' wage recently has been running at about $40,000 a year, which is equal to unionized steelworkers' pay. Unlike a union wage, the Nucor wage is more bonus than anything else. "More than half of the steelworkers' total compensation comes from production bonuses," said Iverson.

All of the Nucor factories had been visited by labor organizers and none of the factories had mustered the initial vote needed to start a union election. When a union organizer showed up at the Darlington steel mill, a delegation of steelworkers went to the mill's general manager and said to him, "What did you want us to do to the guy?"

"Don't do anything to the guy!" was the manager's answer. He ordered a foreman to accompany the union man around the mill, to prevent the union man from being made into a bar of steel by the Nucor steelworkers. There is another way to look at this story. It was pretty clear that if a Nucor plant voted to join a labor union, Iverson might decide to sell off the plant or even close it down. As Keith Busse expressed it, "If the unions took over one of our plants we would turn as mean as a junkyard dog."

"We're against labor unions," said Iverson. "It's not the union pay scale that we object to, it's the work rules." He believed that Nucor employees were not likely to join a union, anyway. "Some of our people have been exposed to union jobs," he said. "They've been laid off or gone out on strike. They associate unions with strikes and layoffs. They want security for their families. The union doesn't represent security to them, but just the opposite. We don't hesitate to point out to our people that there were once 450,000 steelworkers in this country and now there are 130,000, and that happened at the height of the unions. What can the union bring to our people? I think there are companies in the United States that need unions. Companies need unions when management doesn't know how to set up progressive programs and run them. But the unions lost sight of the

■

fact that in order for employees to do well the company has to do well. The unions are caught in a bind over work rules. They can't get rid of work rules. If you get rid of work rules, then you reduce the number of people working in a plant, and productivity goes up, and then fewer people can share more of the success of the company. But then the unions lose membership, because there are fewer people working in the plant. So the unions can't abandon work rules, because that would threaten their existence.''

■ ■ ■

The Nucor workers earned pay that was double or triple the average laborer's pay in the small towns where Nucor built its mills. Nucor gave its employees their bonuses every week, so that they did not have to wait around for the end of the year to see what kind of spare change the management might chuck in their direction, after the management had decided upon its own bonuses. Iverson gave the Nucor production workers a steady ten percent of the Nucor Corporation's pretax profits every year for their retirement plans. Since the company's profits varied from year to year, the steelworkers had no idea how much money they would be worth when they retired. When an employee retired, Nucor paid him his retirement savings in a lump sum; no pension.

Some Nucor steelworkers, who had been with the company since the Darlington startup, now had more than $200,000 in the bank, sitting in their individual Nucor retirement plans, waiting to be drawn as a single check when they retired, and a portion of that wealth was laid up in the form of Nucor stock. If Nucor's stock were to grow rapidly in value, some of the long-time employees might get a check for $500,000 when they retired. On the other hand, if the Crawfordsville Project failed and Nucor's stock dropped in value, then every steelworker at the company would have a diminished nest egg.

Nucor production workers received college tuition money for their children, $1,800 a year for each child in college. Iverson gave all Nucor employees stock in the company at regular intervals. He did not lay anybody off, although he never promised not to lay people off. Iverson had managed to get through twenty-five years in the steel business without a layoff. Except once, when a Nucor general manager laid off forty Nucor workers. Iverson ordered the manager to rehire the workers, and soon afterward he fired the manager. The message was not lost on the company's managers. There has not been a layoff at Nucor since then.

Nucor's steelworkers work in production teams of about thirty people,

■

all of whom are responsible for the efficiency of the team. The rules are strict, if not ruthless. The pressure to keep up production comes from the rules, and the rules are enforced by the production team itself. If one person in the team is late for work or doesn't show up for work, the paychecks of all thirty people in the team can be damaged that week. The reaction to absenteeism at Nucor is swift and unpleasant, and comes from other members of the production team rather than from managers. People who become sick may try to struggle through a day's work for fear of hurting or irritating their fellow team members. If a person on a team is thought to be lazy by the others, they nag him and lecture him, and if he doesn't take to lectures, they get him fired.

"These are a tough people," said Iverson. "You have to be tough to make steel. Making steel is hard, dirty, dangerous, skilled work." In the early years the teams had all been male, but lately one had begun to see hot metal women in the production teams, here and there.

During a factory startup, the production teams gradually take control of the factory. The teams themselves begin to decide who shall stay with the team and who shall be fired. Plenty of people are fired, or they hate the work and quit in frustration. Employees who don't like the pressure of a Nucor plant startup don't stay with Nucor. It is a Darwinian law of non-union steel, in which the rewards are money, but the work is not all that safe, and the people who can't handle the job are many. "If there's one guy in a group who isn't doing well," said Iverson, "the others in the group either train him or get rid of him." One time, some members of a team in a Nucor Vulcraft joist plant chased a guy around the plant with an angle iron. They figured that the quickest way to get him off the team was to kill him.

■ ■ ■

Nucor's factory operators were known as vice-president–general managers. They were officers of the corporation and they had been given wide executive powers. They worked at a Nucor factory and ran the factory. They were responsible for the plant's finances and accounting, labor relations, employee safety, product development, advertising, and marketing.

"In many ways they are like presidents of their own operations," said Iverson. "I let the general managers rattle around in their own cages. Some of them are a little bit like dictators."

Nucor's general managers were a collection of field generals, with wide authority to run their war as they saw fit. Most of them were engineers by training; Keith Busse, with an accounting background, was an exception. In reality, some of Nucor's factories were coasters, not making money and not losing money, merely generating a little bit of positive cash flow. Rarely if ever did all of Nucor's factories earn a handsome profit in the same year, and consequently Nucor was one of those companies that often fell a little below Wall Street's hopes, and so the stock went through the occasional vicious sell-off.

Iverson gave the general managers bonuses that were tied to the corporation's profits, and when the profits fell, the managers were hit hard. "I call it Nucor's share-the-pain program," as he put it. Iverson paid himself a salary that was about seventy percent bonus, in both cash and stock. Iverson calculated his bonus according to Nucor's profits. In recent years his total compensation had gone all over the place, varying between $108,000 and $753,000. In a boom year Iverson's pay hit the middle range of pay in the Forbes 800 list of the salaries of chief executives in America. In a bad year Iverson's pay hit the skids, dropping close to number 800 on the Forbes list, right at rock bottom. Iverson punished himself for poor earnings at Nucor, an unusual practice among top executives.

Nucor had only three top executives: Ken Iverson, the chairman and CEO; David Aycock, the president and chief operating officer; and Samuel Siegel, the chief financial officer. Those three executives also comprised the company's board of directors, along with one retired Nucor executive.

David Aycock, a thickset man with a square jaw, whitish blond hair, and a belly, has a high school diploma and is worth $15 million to $20 million. He grew up on a farm in Anson County, South Carolina, where his father raised cotton on twenty-five acres, plowing the land with mules. When he was a boy, Aycock didn't think that his family was poor: "Hell, we had two mules, we were doin' all right." He did not always agree with everything that Ken Iverson said to the press. "I doubt that U.S. Steel and Bethlehem Steel are scared of us, like Kin says," said Aycock. He called Ken Iverson "Kin." "I say I doubt it because I've never been scared of them."

"Are you worried that the Crawfordsville Project might fail?" I asked David Aycock.

"I don't *choose* to be worried," he replied, in a soft Carolina voice.

Aycock's office is two doors down the hall from Iverson's office. When Iverson talks on the telephone, Iverson's words are billiard balls whacking along the hall into Aycock's office: "Hello! Ken Iverson here, Nucor! Iverson! Spell it? Sure. I-v-e-r-s-o-n. Nucor! Nuke Ore!"

■

"Kin may have learned how to talk in a sawmill," said Aycock.

When Keith Earl Busse, vice-president–general manager of the Crawfordsville Project, is on a visit to headquarters, the loudness between Busse and Iverson can get loud enough to flood the lobby of Nucor's rented suite with noise. In the lobby, Nucor's customers and vendors are sitting on motel chairs, nervously crossing and uncrossing their legs, unable to ignore a voice saying, "Ken, don't give me that shit!" and Iverson's voice booming, "God damn it, Keith!"

The customers and vendors glance at each other. Is Mr. Iverson firing this Keith guy? Suddenly they hear footsteps running in the hall—what is this, a fistfight?—no, it's Iverson's secretary hurrying to close Iverson's door. The receptionist smiles brightly and says, "Oh, that's just Ken talking with Keith."

■ ■ ■

The modern Nucor was in many ways a reaction against the operating style of David Thomas, the former chairman of Nuclear. The modern Nucor had no airplanes, no executive perks. In every detail of Nucor's structure, Iverson tried to keep things informal and simple. Iverson had a secretary, but he answered his own telephone, because his secretary had better things to do with her time than to take the chairman's calls. It is possible that Iverson is the only CEO of a Fortune 500 company who answers his own telephone.

If you wanted to talk to Iverson, you rang up Nucor's headquarters on the phone and you got, "Good afternoon, Nucor"—the receptionist by the starveling umbrella tree—and next you got, "Hello, Ken Iverson." When he talked about Big Steel he sounded puzzled and surprised and even amused. He portrayed the Big Steel companies as victims of their instincts, unable to change, unable to think, doomed by their anatomy. "The Big Steel companies tend to *resist* new technologies as long as they can," he said. "They only *accept* a new technology when they need it to *survive*. If any of the Big Steel companies *did* decide to build a mill like the one we're doing in Crawfordsville, it would take them two or three years to finish it, at least. And it would cost twice as much as ours. In any case, the Big Steel companies don't have a whole lot of capital right now. The banks won't lend 'em money."

"That's terrible," I said.

"Oh, I never fault a bank for being *conservative*." A conspiratorial laugh, *huh, huh, huh*. "The Big Steel companies got themselves into that

■

position. Because they didn't spend their money wisely. Since the end of World War II, they have not been willing to accept new technologies. They kept trying to *upgrade* their plants to *older* technologies that everybody else in the world already used. The Big Steel companies have their own in-house engineering departments. They have their own in-house *consulting* departments! A project is designed by a large committee of engineers and consultants. There can be as many as fifteen people on the committee that does the engineering work for a project at Big Steel. Then the project goes to an executive committee for approval. Then the executive committee sends it *back* to the engineers. These monstrous bureaucracies are unwieldy, to say the least. Don't get me wrong in my criticisms of Big Steel. My argument is with all of corporate America.''

David Aycock put it like this: "If you see an organizational chart with lines of authority runnin' every which way, you'd better believe it's a disorganized company.''

The usual practice at a growing company would be to build a corporate headquarters. There are two reasons for a corporate headquarters: it serves as a visible monument to the company's success, and it provides office space for managers. For those same two reasons Iverson never built a corporate headquarters. He resisted moving the corporate headquarters out of the rented suite in the Cotswold Building, because he was afraid of self-congratulation and because he didn't want any more managers at the company. He literally slammed the door on management.

Chief executives of Fortune 500 companies sometimes dialed Nucor's headquarters, seeking help. They wanted Iverson's advice on how to hose out a bureaucracy. As Iverson put it, "They say to me, 'Ken, I know I've gotta do it, I know I've gotta clean out these layers of management, but there are all these people here that I've worked with all my life. I have to fire them, and they are my friends. I can't do it. I can't do it!' '' Iverson did not know what to say to the lonely executives who called him on the telephone, because he believed that once a termite nest is embedded within a corporation, getting rid of the nest is going to be painful and difficult, involving a great deal of suffering.

Iverson was a popular speaker on the Rotary Club circuit, where he lectured on failure as the normal and expected result of management. One of the talks he gave to Rotary Clubs was entitled "Good Managers Make Bad Decisions.'' He informed the Rotarians that human nature is fallible. "The best manager in the world, a guy with a Harvard M.B.A., might make bad decisions around forty percent of the time,'' he said. "And a

■

rotten manager might make bad decisions *sixty* percent of the time.'' Iverson suggested to the Rotarians that there was approximately a twenty percent difference in performance between a rotten manager and a Harvard M.B.A. ''As a manager,'' said Iverson, ''you have to make decisions. If you don't make decisions, you are going nowhere and doing nothing. But if you make decisions you will make bad decisions. You have to have a strange and monstrous ego to think that you never make bad decisions. We tell our employees that we do make bad decisions.''

Two business professors from the University of Massachusetts once arrived at the Cotswold Building to study the management structure of the Nucor Corporation. They found an inadequate management structure, in their opinion, and they felt moved to speak. In hushed voices, behind a closed door in Iverson's office, the professors told Iverson that Nucor couldn't be managed properly with a handful of people working out of a rented suite. Iverson replied that Nucor could double in size without putting a strain on the corporate staff. The professors went away a little bit baffled.

Nucor's visible lack of management is still a management, and Iverson claims that it's a normal management. ''Unorthodox? It's basically common sense. The way some companies are operated in this country, they are far more unorthodox than Nucor.''

David Aycock put it this way: ''You can't manage people. You *can* bribe 'em.''

In Iverson's view, an authoritarian, hierarchical management creates a need for labor unions. ''When there's a deep-seated conflict between management and labor, it's because of autocratic management practices,'' he said. ''A manager says to himself, 'I'm going to make the guy do it.' Not, 'I'm going to help him do his job.' You can't *make* a person do something.''

Said Aycock, ''If you could get into your employees' minds, you could manage 'em, but you can't get into their minds. People are free in their minds, and you can't manage a free mind. The main thing is to practice leadership. If I could do a perfect job managing myself, then I could lead people.''

That is easier said than done, because to quit managing is something that managers don't like to do. As Keith Busse put it, ''In many respects Ken Iverson is as common as an old shoe. Ken never flaunts his money or his power. Well, he never flaunts his money. But Ken can come down on your neck with the sheer brute power of his authority, when he wants to. Most

■

of the world conceives of Ken Iverson as a hands-off manager, but that is not strictly true. Ken wants to stick his fingers into things and get involved with them. He preaches differently, but he will dictate to people. He will step into the fray several levels down, and no chairman should do that. Do I shout at Ken Iverson? My voice goes up a few decibels, if you want to call that shouting. I say to him, 'God damn it, Ken, that will not work!' "

■ ■ ■

Donald C. Lillis, the investment banker who had amassed twenty-four percent of Nuclear's shares when he bought them from Martin Marietta for a nickel apiece, remained a director of Nuclear for a couple of years after Iverson became president. One year at a shareholders' meeting, a shareholder stood up and asked Lillis if he would ever consider selling off his Nuclear shares.

"I don't plan on selling my Nuclear shares," replied Lillis. "But I've been on Wall Street for fifty years, and if there's one thing I've learned, it's that at the right price I'd sell the shirt off my back." Soon afterward Lillis died of a heart attack, and so God parted Lillis from his Nuclear shares. Lillis's executors and family evidently believed that the dead man's Nuclear was a rank speculation, and sold off Lillis's stock to the general public for the equivalent of $1.50 a share. Since then, it has split 2.2 for 1 and now trades at $63 a share—as of this writing. It has gone up 280,000 percent over what Lillis paid for it. These old Wall Street sharks never get the respect they deserve from their offspring. If the Lillis family had hung onto the old man's Nuclear, the stake would be worth at least $200 million, with a cost basis of $70,000.

Recently Sam Siegel hired Goldman, Sachs & Company, Nucor's investment bank, to manage an auction of Nucor's Research Chemicals division. Research Chemicals was the little division that had now been searching for thirty years to find uses for "the 14 comparatively unknown elements known as rare earths," in the words of the visionary Sam Norris, the first president of the Nuclear Corporation of America. Norris's vision paid off. Goldman, Sachs knocked down the rare-earth division for $79 million in cash to Rhône-Poulenc, the French chemical giant. At irregular intervals Siegel gets a letter from someone wondering if his or her stock certificate in Reo Motors, Inc., is worth anything. Siegel converts the Reo shares to a large number of Nucor shares, which makes the claimant happy. At very rare intervals, Iverson gets a letter in broken English from someone in Latin America, wondering if Mr. F. Kenneth Iverson will sell

■

him something like a transmission for a 1939 Reo Speedwagon delivery truck that just blew up. America's toughest trucks seem to be holding up well in Latin America. Iverson writes back to say that he doesn't have any Reo parts and, unfortunately, he doesn't know where to get any. He doesn't mention the Reo hubcap that sits on the shelf near his desk.

7

THE
IMAGINARY
MACHINE

■ ■ ■

The top Nucor executives and board of directors—Iverson, Aycock, and Siegel—met nearly every day at Phil's Deli, in the Cotswold Shopping Center. They referred to Phil's Deli as "the executive dining room." They regularly ate lunch at Phil's, and these meetings were about the only time when they met as a group to talk about the steel business. Most of the company's important decisions were made at Phil's Deli over sandwiches. Iverson preferred the cream cheese and olive sandwich and Aycock preferred the grilled cheese sandwich. Sam Siegel really preferred to eat Chinese food at the Hunan Palace across the parking lot, but Siegel's garlic powder had cured Iverson so permanently of garlic that Iverson absolutely refused to eat at the Hunan Palace with Siegel. Huddled at a bistro table at Phil's, the three executives plotted hundred-million-dollar guerrilla raids against Big Steel in voices muffled by cheese.

The executives were not unaware of the fact that Nucor was trapped in the borderland between bar steel and greatness, and in recent years the question at Phil's Deli had been how to grow Nucor into one of the world's great steel corporations. At the center of Nucor's hopes, there was a Machine that would cast steel into a thin, flat strip. This imaginary Machine would enable Nucor to manufacture sheet steel at a low cost of capital. The Machine would bypass the high cost of rolling machines, and rolling machines are the most expensive part of a steel mill. The only problem was that the Machine did not exist.

Sir Henry Bessemer, an English inventor who lived during the reign of

■

Queen Victoria, first saw the possibilities of the Machine. (Bessemer is known for his invention of the Bessemer converter, a jug-shaped furnace that converted liquid pig iron into steel by the ton, cheaply.) In 1856, the same year in which he invented the Bessemer converter, Bessemer built and patented a strip-casting machine for the direct casting of molten steel into a ribbon.

Bessemer's machine contained a funnel—which today is more properly called a mold—which gathered a flow of liquid steel and delivered it vertically downward between a pair of steel rolls, where the steel hardened into a thin strip, and then the strip was withdrawn from the machine in an endless ribbon. It was a simple solution to what looked like a simple problem. But when Bessemer poured steel into his machine, the machine welded itself together. He abandoned the experiment.

The second attempt came in 1891, when Edwin Norton, of the Fluid Metal Rolling Company in Maywood, Illinois, built a machine that featured a pair of giant steel rolls, eight feet in diameter, between which molten steel was cast into a ribbon. Norton made several tons of steel ribbons but ran out of money. For the next hundred years the inventors kept trying and trying, and the busts went on and on, it being the nature of liquid steel to ravage the cuckoo clocks of inventors.

The chronicle of hot metal nightmares began for real in 1921, when an American inventor, Clarence W. Hazelett, began a fourteen-year series of experiments with a twin-roll strip-casting machine (U.S.A., 1921–1935; a success with liquid brass, a bomb with liquid steel). Undeterred, Clarence Hazelett invented another steel-casting machine, the Hazelett ring caster (U.S.A., 1935–1939; a failure). Also there was Professor Ulitovsky's strip-casting machine (U.S.S.R., 1935–1939; a failure). There was the Goldobin machine (U.S.S.R., 1938; a failure). There was Mr. J. M. Merle's machine (U.S.A., the thirties; a failure). Near the end of his life, Clarence Hazelett got back in the game with a twin-belt steel-casting machine (U.S.A., 1954; "promising," but it ended with Mr. Hazelett's departure into a more promising world). Hazelett's son, R. William "Bill" Hazelett, carried on the old man's work with twin-belt steel-casting machines (U.S.A., Japan, and France, 1964–1969; five failures). There was the Creusot-Loire machine (France, 1972; abandoned). There was the American Can thin-strip steelmaking machine (U.S.A., 1972–1980; "phased out" by skeptical management). There was the LTV ring-casting machine (U.S.A., begun in 1963, abandoned in 1975, revived in 1985,

■

abandoned in 1986 after a total expenditure of $300 million). There was the first British Steel channel caster (Great Britain, 1975–1979; abandoned). There was the *second* British Steel channel caster (1981–present day; clinically dead). There was the Hitachi-Korf wheel and belt machine (Japan and Austria, 1980–1985; "in development"). There was the Sheffield Peeler (Great Britain, 1982; weird; it "peeled" a solid ingot into a thin strip, the way a log is peeled into a sheet of plywood). There was the Osprey machine (worldwide, 1976–present day; nothing that quite works). There was the Kawasaki vertical belt caster (Japan, 1983–1989; died amid puddles of leaked steel). There is the Grant machine at the Massachusetts Institute of Technology (U.S.A., present day; success claimed). There is the Irsid/Clecim machine (France, present day; comatose). There is the Nippon drag caster (Japan, present day; dragging in every sense of the word). There is the Concast "uphill" machine (Switzerland, never built; works smoothly on paper). Finally, there is the Argonne electromagnetic levitation process (U.S.A., 1984–present day; the machine is supposed to hold liquid steel suspended in midair by means of magnetic fields, but so far it has managed to levitate only one kilo of hot metal).

The inventors were troubled by a recurrent nightmare. The nightmare concerned the fact that most of a steel-casting machine's parts are themselves made of steel. Therefore the machine is apt to weld itself into a lump. It is like trying to manufacture ice by pouring water into a machine made of ice—freeze-ups and meltdowns are inevitable. A thin strip of steel hurries quickly through the machine, and rapid movement requires complicated machinery, and complicated machinery can be destroyed by hideous splashes of liquid steel. Spilled steel can start fires. Steel fires have a queer way of turning into electrical fires, because when steel gets loose it runs everywhere, melting electrical cables. Liquid steel tends to form clots and icicles inside a machine—they are called skulls—and the skulls can lodge between moving parts, giving the machine a heart attack.

Molten steel will erode any material that it touches. The container has not yet been invented that will hold steel in a hot, fluid state for a reasonable period of time without being eaten away. Liquid steel has a frightening tendency to burst through containment vessels. No material has ever been discovered that can serve as a container for liquid steel without being eroded, burned, or dissolved, not even Carborundum, graphite, tungsten, zircon, ruby, or fused sapphire. Diamond won't work either—since it's pure carbon, steel dissolves it. Running steel cuts gullies in ceramic firebrick the way running water cuts through soil. Liquid steel is chemically reactive. Depending on what substances are dissolved in liquid steel, the metal can act as an acid or as a powerful caustic—liquid Drano at Fahrenheit three thousand.

■ ■ ■

The problem of how to cast a continuous strand of steel was partly solved by the invention of the so-called "classical" continuous-casting machine, which can cast a square or rectangular strand of steel (a billet or a thick slab). This major achievement was pioneered during the twenties by a German inventor, Siegfried Junghans, and was brought to commercial success in the early fifties by an American gadgeteer, Irving Rossi. The Junghans-Rossi classical continuous-casting machine is the prevailing technology for making a solid piece of steel.

The secret of the classical Junghans-Rossi caster is a vertical box-shaped funnel, or mold, that oscillates up and down. Liquid steel is poured into the funnel, and as the funnel oscillates, a rectangular piece of steel is withdrawn downward from the bottom of the funnel. The slab moves very slowly through the machine, being cooled with huge amounts of water. As the solid metal is extruded, it is cut into lengths. The slabs are then rolled through a rolling mill, into sheet steel.

The problem with the classical steel-casting machine is that it extrudes a slab that is as thick and wide as a queen-sized mattress and is about as long as a driveway. In order to crush a fat slab of steel into a thin sheet, it must be passed through a train of millstands half a mile long, a hot-strip rolling mill. It costs $1 billion to build a hot-strip rolling mill capable of crushing a steel mattress into a sheet. That is why only Big Steel companies make sheet steel.

■ ■ ■

The top managers of the Nucor Corporation discussed their plans around a bistro table at Phil's Deli in low voices, wiping crumbs off their lips with paper napkins. In 1986, they committed $5 million to purchase an experimental steel-casting machine built by the Hazelett Strip-Casting Corporation of Colchester, Vermont, a private company headed by R. William "Bill" Hazelett.

At that time, the whole world was playing with Hazelett machines. In Japan, Sumitomo Metal Industries had put a swarm of thirty researchers to work on a Hazelett machine to try to persuade the thing to make steel.

■

Meanwhile, U.S. Steel and Bethlehem Steel had got a government grant from the U.S. Department of Energy for a joint research project into Hazelett machines. The two Big Steel companies put to work a total of 150 scientists and engineers on two Hazelett machines.

Nucor's Hazelett machine ended up in the hands of Mark Millett at the Nucor minimill in Darlington, South Carolina, where Millett was working as a metallurgist. In David Aycock's words, "When we installed the Hazelett machine at Darlington, we knew one thing. We knew we had to turn it over to very young people. It never crossed their minds that the machine possibly might not work." Millett was then twenty-six years old.

Millett was given one full-time assistant (a certain Bill McKenzie, a foreman), a half-time electrician (one Richard Brewer), and a "laboratory," an empty room near the furnaces. Millett had almost no experience with steel-casting machines. It was a nerve-racking assignment for Millett, since Japanese and American steel companies had committed a total of 180 scientists and technicians to swarm over Hazelett machines, as if the machines were about to be launched into orbit.

The Hazelett machine was a Yankee contraption from Vermont. It worked like this: you were supposed to pour molten steel between a pair of water-cooled conveyor belts, which were made of flexible sheet steel. The steel solidified between the belts as the belts carried the steel along, and then the belts separated and peeled away from the steel, like waxed paper being peeled off a slice of liverwurst.

Millett read up on the history of failures with Hazelett casting machines. The machine had been invented in the fifties by Clarence W. Hazelett. It had shown much promise, but during the sixties no less than five experimental Hazelett machines had bombed, oozing out steel that looked like beef jerky. The machines had failed because a nozzle that was used to inject liquid steel into the machine tended to swirl with turbulence. As the liquid metal entered the mouth of the machine it tumbled and rippled, and that caused the machine to belch, vomit, or to jam up with lumps of solid steel, and the steel that came out of the machine was covered with scabs of black slag. Millett began to consider ways to gently ease molten metal down the throat of the machine without turbulence, so that a smooth slab of metal would come out of the machine.

"I did some thinking and I came up with a variety of ideas for nozzles," he said. They were Millett nozzles. With glue and large pieces of plastic, he built full-sized models of various Millett nozzles in his laboratory— weird flaring objects. Now he needed water. He ran a hose from an outdoor

■

faucet in through a window of his laboratory and poured water through the plastic nozzles, watching the water ripple and burl. Visitors to Millett's laboratory usually found water all over the floor. Eventually Millett came up with a promising nozzle.

"It was an unconventional, three-cornered spout, like a spout on a teapot," as Millett describes it. Having grown up in England, he was familiar with the spout of a teapot. He next built a Millett nozzle made of ceramic material, strong enough to hang together when molten steel ran through it. I can't describe the exact shape of the Millett nozzle, because I don't know exactly what it looks like. Millett is vague when he describes the shape of the nozzle that he invented. The shape of the Millett nozzle is a secret owned by the Nucor Corporation.

The Hazelett machine had been placed in a corner of the Darlington melt shop. Millett fitted a Millett nozzle into the lips of the Hazelett machine and then he called for hot metal. A couple of steelworkers brought over a ladle of molten steel with a crane, and they poured the steel into a ceramic holding tank. The tank filled up with hot metal, and the metal ran into a second tank—complicated plumbing—and from there the steel burbled through the Millett nozzle into the Hazelett casting machine. It entered the throat of the machine. Millett mentally crossed his fingers. Wheels turned and belts moved, and steel began to move through the machine. Silence.

"What's this?" Millett said to his assistant.

Suddenly a bright light streamed out of the machine, and there was a crackling hiss, like a grease fire.

"Ahoy! What's happening?" Millett said.

The machine couldn't take it. It let out a farty rush of smoke and then regurgitated steel right back through its mouth. The steel bubbled out of the machine's lips and back around the Millett nozzle and poured to the ground with a horrible frying sound, a sound like an institutional deep-fat cooker doing a bulk job. It made a sound like a bucket of chicken parts being dropped into a Pitco Fryalator.

Millett lowered his dark glasses to protect his eyes from the steel fire burning on the floor and from the glare of the fizzling machine, and backed slowly away from the machine.

They let it cool and then cleaned it up with crowbars and cold chisels, and, a few days later, Millett tried again. This time David Aycock, the president of the company, drove down to Darlington to see what Millett was doing. Millett had made a few changes in his nozzle and the plumbing, and he and Aycock stood before the machine as the hot metal ran through the Millett nozzle into the machine. Wheel and belts turned, there was a pause, and then ten feet of glowing solid metal was rapidly extruded from the hindquarters of the machine.

"Hm," said Millett.

Suddenly there was a violent hiss, a burst of light, and liquid steel fizzed and puked from the mouth of the machine. The machine burst into flames and vanished in a cloud of steam, dropping cow patties of steel to the floor, and Millett and the president backed away from the breakout to save their skins.

"We did the research in the Nucor style, which is to get down and dirty and get the job done," as Millett expressed it.

Just before Christmas of 1986, he poured a heat of steel into the Hazelett machine, and out of it came a long thin slab of smooth commercial-grade steel.

"Excellent!" Millett remarked.

Millett's assistant, Bill McKenzie, later commented, "That was good steel. You could have pounded garbage cans out of it all day."

Millett continued to make commercial-grade steel. According to some reports, Millett's steel was consistently better in quality than the steel that was coming out of Hazelett machines operated by the Japanese and Big Steel research groups. In Japan, the Hazelett experiments drifted off to sleep. In the United States, U.S. Steel and Bethlehem Steel, after spending $20 million of taxpayers' money on research, declared the Hazelett machine to be essentially a failure, and aborted their project. Mark Millett had done at least as well as twenty million federal dollars, a zillion yen, and the combined intelligence of 180 Japanese and American experts in the art of casting steel.

■ ■ ■

On March 18, 1983, in the city of Düsseldorf, West Germany, an obscure engineer by the name of Manfred Kolakowski had a sudden idea for a steel-casting machine. Kolakowski was an employee of SMS Schloemann-Siemag A.G., a machinery firm headquartered in Düsseldorf. He saw in his mind's eye the Machine. Rumor has it that Kolakowski was sitting on the toilet at the time.

"No, no, no, no! I was not sitting on the toilet! I was sitting at my desk!" the inventor told me. "I was talking on the telephone! And do you know how one will draw little pictures of little men and so forth when one is talking on the telephone? I drew a little picture, while I was talking on

■

the telephone. I can swear on it! This story that I was on the toilet, that's not right! The newspapers in Germany will try to make exciting stories out of a thing like this!''

One of Kolakowski's friends offered a different version. The friend's unauthorized version appeared some years later, after Kolakowski's machine had been built. ''Manfred and I were sitting around one day drinking coffee, after we had just put a full ladle of steel through his machine, and we were feeling good about it. So we were sitting there drinking coffee, and I said to him, 'Manfred, how in the hell did you really dream this machine up?' And he got this little smirk on his face, and he said, 'Honestly, to tell you the truth, I was sitting on the toilet.' ''

In any case, Manfred Kolakowski had an idea, on the eighteenth of March, 1983, that changed the world steel industry. In German the word for a sudden idea is *Einfall*. It means a falling in.

What fell into Kolakowski's mind was a vision of a three-dimensional shape. It was a funnel for casting liquid steel, technically known as a mold. It had curved, bulging sides, and it narrowed down into a slit. In fact, the Kolakowski mold did vaguely resemble a toilet. It gathered a flow of liquid steel and narrowed it, and sent the steel downward in a sheet, wherein the steel solidified. The Kolakowski mold, or funnel, looked like a letter envelope that has been squeezed open. The funnel was narrow and flat, but it bulged in the center. It was a recurvate lens, with a swollen core and thin, bladelike edges, like a melon seed. It was to be the heart of a steel-casting machine. The funnel, or mold, would cast steel vertically downward through its slit into a train of rolls, where the steel would harden into a thin slab, and the slab would be bent and withdrawn horizontally from the machine in an endless ribbon.

Kolakowski sketched his idea on a piece of paper and handed the sketch to a fellow engineer at SMS Schloemann-Siemag A.G. This engineer's name was Hans Streubel. Streubel began to think about the mold, and in talks with Kolakowski, Streubel made significant improvements in the design. Their employer, SMS, took out international patents on the copper funnel, and Manfred Kolakowski and Hans Streubel shared the patent award, fifty-fifty. I will call the funnel the Kolakowski mold, because the first vision of the design came to Kolakowski, either at his desk or on the hopper, but like many inventions it was a collaborative effort. This is particularly true in a German engineering firm such as SMS, which is staffed with many engineers who work closely together on common projects.

■

SMS operates a foundry in the Siegerland, a mountainous region of north-central Germany, where secret experiments can be conducted. Kolakowski and Streubel made a funnel out of cast iron in the Siegerland foundry. The funnel, or mold, had a slit in the bottom. Next they tried an experiment: they installed a plug in the bottom slit of the funnel. They attached a hydraulic ram to the plug. Then they filled the funnel with molten steel, and that was more or less like a toilet filling up. Then they flipped a switch and the hydraulic ram extracted the plug—they flushed the toilet. There was a cracking noise, and as the plug came out of the funnel, a thin, orange-hot, smooth, rectangular strip of solid steel came out of the funnel, stuck to the plug. They did it again: they filled the funnel with molten steel, hit the switch, and the ram pulled the plug and dragged out a thin slab. It was mysterious to watch, it was eerie, it went against common sense. It was like seeing a magician pour a quart of milk into a rolled-up newspaper and drag out a long white handkerchief.

When SMS executives saw Manfred Kolakowski do his trick with molten steel, they were impressed. SMS spent $7 million constructing a prototype of a Kolakowski-Streubel casting machine at the Siegerland foundry—a pilot machine, complete with rolls and motors and computers, the whole works. SMS put Manfred Kolakowski in charge of the pilot machine. SMS shrouded the project in secrecy, fearful that its competitors would steal the design.

In the privacy of SMS's hot metal laboratory in the Siegerland, Manfred Kolakowski poured hot metal into his pilot machine. He got a mushroom cloud, a flash of flame, an earsplitting roar—a breakout. Undeterred, he began a series of what he called "campaigns," pouring different types of steel into the machine to see what would happen. Often he got a breakout: torrents of liquid steel rushed through the machinery and a siren went off, *whoop, whoop, whoop!* while German steelworkers scurried around with the impotent fury of ants coming out of a kicked nest, yelling, *"Breakout! Breakout! Wir haben einen Breakout! Das Scheißding hat einen Breakout!"*—"Breakout! Breakout! We're having a breakout! The shit-thing had a breakout!" Among steelmakers the word "breakout" transcends earthly tongues.

Kolakowski's campaigns drove on through breakouts, until by 1986 his machine was extruding thin slabs of carbon steel. The slabs were 1½ or 2 inches thick, and thus easily rollable into sheets. SMS engineers got busy and drew up plans to link the casting machine directly to a train of four millstands, a miniature hot-strip rolling mill that would crush the glowing

■

slabs down to sheet steel and shoot out metal like Saran Wrap. It was an integrated manufacturing facility, a high-speed continuous steelmaking machine. Two SMS design teams, under the leadership of Heinz Scholz and Max Münker, tied the parts together into one machine. And so was born the Compact Strip Production plant, a design for the hottest racing engine in the steel industry. It ran beautifully on paper.

At SMS, they began calling Manfred Kolakowski the Pope of the Strip Caster. All visitors to the SMS foundry in the Siegerland had to sign a secrecy agreement before viewing the pilot machine or even speaking with Kolakowski. A total of three hundred executives and engineers from at least a hundred different steel companies visited the foundry, signed the secrecy agreement, and watched Manfred Kolakowski make thin slabs of steel. Everyone visited the SMS foundry in the Siegerland—the Japanese, the Koreans, the Brazilians, the Taiwanese, the Australians, as well as executives from virtually every major American steel company. None of the steel executives, not the Japanese, not the Koreans, not anybody, wanted to buy a CSP. The steel executives looked but they didn't want to touch.

There were two reasons why they didn't want to buy a CSP. In the first place, the CSP might fail. In the second place, it might work. Either possibility dismayed steel executives. If the machine failed, it would leave a wound in someone's balance sheet totaling up to a half a billion dollars in write-downs. If it worked, then pretty soon any moron with half a billion dollars could get into the sheet steel business by building a desktop steel mill. The CSP would render large steel mills obsolete. The whole idea of a cheap, miniature, continuous-steelmaking machine was deeply threatening to large steel companies. It threatened their domination of sheet steel markets. It threatened their financial health; it threatened their way of life.

■ ■ ■

In the spring of 1986, David Aycock, the president of Nucor, traveled to West Germany to inspect Kolakowski's pilot machine. "That son of a gun is going to work," thought Aycock, as he watched a strip of red-hot steel rumble out of it. He returned to Charlotte and told Ken Iverson that the son of a gun was going to work. Iverson began a series of secret meetings with the German company. On December 7, 1986, a couple of weeks before Mark Millett began to produce good commercial steel with a Hazelett machine, a group of SMS executives flew into Charlotte, North Carolina,

■

for another secret meeting with Iverson. The Germans must have thought that something big was coming, because the delegation was led by a certain Hans-Friedrich Marten, a member of the managing board of directors of SMS. The Germans settled themselves on a fake-suede couch in Iverson's office and piled their blueprints on a coffee table, whereupon Iverson told them that Nucor was ready to proceed immediately with a steel mill containing a Compact Strip Production machine.

The Germans were stunned. They had expected the negotiations to continue for at least a year. They smelled a contract in the air, and they went for it.

"It was a grueling, nerve-racking session," Iverson later recalled. "Mr. Marten and I were both trying to get the best deal we could for our companies. We had a lunch brought in from Phil's Deli." Lights burned late in the top corner of the Cotswold Building, and the Germans stayed up all night in their hotel rooms at the Park Hotel in Charlotte, drawing up a handwritten proposal for a steel mill. The deal was concluded the next morning at the Park Hotel on a handshake between Iverson and Marten, and they signed a letter of intent agreeing to build a steel mill. The Germans were in a state of shock. These Nucor people moved so fast. They drank one bottle of chilled white wine with the Nucor people, to celebrate and cool off. Then Hans-Friedrich Marten instructed one of the German executives to drive him immediately to the airport, and Marten caught the next plane back to Germany, evidently in a big hurry to get out of Charlotte before Iverson changed his mind.

Iverson would not tell me how much he had paid for the Continuous Strip Production machine, but I gathered that the price came to around $70 million, or somewhat less than the current price of a decent painting by Vincent van Gogh. Chump change, for steelmaking equipment.

Iverson got a money-back guarantee from the Germans. The Germans agreed that if their machine failed, Iverson would get his chump change back. Iverson, in turn, agreed to pay the Germans royalties from the machine's profits.

Iverson and the Germans also concluded a deal for the machinery for the cold mill (the third building at the Crawfordsville Project). Klöckner Steel, a West German company, had recently closed an antiquated cold-rolling mill in Bavaria, and Nucor, with the help of SMS, purchased the contents of the Klöckner mill for junk, at a price of just $1 million. But it was good machinery made by Vöest-Alpine, the Austrian state manufacturing firm

■

based in Linz, Austria. SMS fixed up the old Austrian machines for another $10 million. Iverson had now lined up the machines for the Crawfordsville Project, and they came from Germany at low, low prices.

Iverson had lost interest in the Hazelett casting machine. At some point—it is not clear exactly when—the Hazelett Corporation asked to see the Millett nozzle. "Hell, no, we weren't going to give it to them, we designed it," as David Aycock put it. Only Nucor possesses examples of the Millett nozzle, hidden under a tarpaulin at the Darlington steel mill. Having committed itself to the German invention, Nucor did not want any American minimills to link the Millett nozzle to a gadget from Vermont and start making the kind of cheap steel that could be pounded into garbage cans. Nucor was afraid of the American minimills. Nucor had begun to look like Big Steel.

8

GERMANY

∎▪∎

The Siegerland, in North Rhine–Westphalia, West Germany, rises into ridges blanketed with fir trees a couple hours' drive east of the Rhine River and the city of Düsseldorf. The Siegerland contains iron. As early as A.D. 900, people of the Siegerland were cutting down trees, burning them to charcoal, and stuffing the charcoal into blast furnaces along with pieces of iron ore; and thus they made the Siegerland into swords. Around 1870, Karl Busse, a carpenter from Siegen, the market town of the Siegerland, emigrated to the United States and settled in Fort Wayne, Indiana. His son, August Busse, married Louise Hanefeldt. August and Louise were Keith Busse's grandparents. Keith can remember Louise chasing him around her dining room table, after he had got his fingers into an apple pie, saying to him, *"Du bist a Dumbkopf!"*—a mixture of German and English. That is all the German that Keith Earl Busse ever learned. Thou be'st a Dumbhead.

∎ ∎ ∎

In June of the year A.D. 1988, Keith Busse and a few of his boys arrived in the Siegerland to inspect the Compact Strip Production machine and conclude some final negotiations with executives of SMS Schloemann-Siemag A.G., the company that invented and manufactured Iverson's machine. It was Busse's first trip to Germany, and he spent some time looking up Busses in the phone book at his hotel. He found altogether too many

∎

103

Busses, but he didn't call any of them because he was too busy buying a steel mill.

SMS has its manufacturing plant in the town of Hilchenbach, near the center of the Siegerland. The SMS plant is a warren of factory buildings that crowd upon Hilchenbach's main road. Forested mountain ridges stand all around the factory. On the day Busse arrived, the mountains were cloaked in mist, and cows were grazing in fields around the factory.

An SMS executive by the name of Franz Küper escorted the Americans into the factory. Küper was a handsome, quiet man in his fifties, with a square jaw, gold-rimmed spectacles, and ash-blond hair painted with one wide splash of gray. He was a specialist in helping underdeveloped countries set up modern steel mills, built by SMS. Küper had done jobs in Brazil, Mexico, Portugal, Argentina, the United States, and other underdeveloped nations. Küper spoke four languages. He was SMS's project supervisor for the steel mill in Indiana. He was the Americans' handler, as it were.

Küper led the Americans through a security gate, through a security door, past some security guards, and into a linked complex of halls filled with German machinery. The machinery was of breathtaking size, and most of it was Nucor's CSP, standing in pieces all over the floor. The CSP was so massive and complicated that it jammed the factory with machinery. One could hardly find a path to walk among these grandiose monuments of industrial power. The Americans picked their way through the machines in a knot, trailed by German engineers. They stopped to admire a forged and gleaming steel gear, ten feet in diameter. It was a transmission gear for a set of rolls in the CSP. The rolls—like rolling pins—would crush the steel flat.

The Nucor people found themselves standing before the CSP's roll housings. They resembled upright chain links. The roll housings stood in pairs. Two roll housings made a millstand. The millstands towered halfway to the roof of the factory.

The Compact Strip Production machine had been built by hand, in the sense that only apprenticed master craftsmen, with years of training, knew how to operate the lathes and machine tools that cut the parts of the CSP. SMS men in blue overalls, master craftsmen, stood on ladders propped against the millstands, and they smiled at Keith Busse and his boys. One man had strung a plumb line against a steel doughnut, a roll housing. He was studying the plumb line to make sure that the millstand stood up straight and true.

■

"This is unbelievable," murmured one of the Americans.

"We are working twenty-four hours a day now," said Franz Küper, as he picked his way among the machines.

"Our rolling mill is of the latest fashion," added a German engineer.

Küper remarked, "I haven't been sleeping lately. It's not that I'm afraid the CSP won't, er, work, but I always get this way before we ship a steel mill to a customer."

"Did you guys forge these things yourselves?" asked a Nucor man, craning his neck at the millstands.

"No," replied Küper. "They were too large for us to forge in our own foundry. But they were forged in Germany. We machined them here in this shop, on a lathe. Come this way, I'll show you the lathe."

The lathe, which had carved the heaviest pieces of the CSP, was a gigantic machine in its own right. At the moment, the lathe was carving a roll housing to a precise shape. The roll housing lay flat on a bed, on the lathe. An SMS craftsman, standing at a control panel, was pushing buttons, while a spinning drill bit the size of a washing machine mixer shaved curls of steel from the roll housing. The curls piled around the roll housing like wood shavings, except they were carbon steel. The lathe had been made by a machinery company in the Siegerland. One began to appreciate the industrial power hidden in the mountains of the Siegerland.

Küper led the Americans into a long hall. At the far end of the hall, a weird instrumentality glittered under floodlights. It was the SMS thin-slab casting machine—*die Maschine*, the pride and hope of SMS, the material result of the idea that had struck Kolakowski at his desk or on the toilet. This was Nucor's casting machine, and soon it would be loaded into an oceangoing freighter bound for America.

The machine was thirty-one feet tall; three stories tall. The apparatus towered to the roof; it was made almost entirely of forged and welded steel; it contained hundreds of very large moving parts; its life-sustaining fluids, watery and hydraulic, were carried in tangles of stainless steel pipes; and it was enameled with gray-green paint. It displayed the SMS symbol:

"Jesus H. Christ," murmured Busse, in admiration.

"It's just amazing how miniature this thing is."

"I feel like a kid with a new toy."

"Funny little thing. I've been around 'em before."

■

Mark Millett was there, and he said to Busse, "So what do you think? Do you think it's going to work?"

Busse grinned and didn't reply.

Ladders and scaffolding had been placed against the machine. On top of the machine there was a box that contained the Kolakowski mold. Below the box extended a double set of forged steel rolls, positioned like the rungs of two ladders that had been placed against each other, beneath the box. As the steel solidified, it would be pulled in a sheet down from a slit in the bottom of the mold and down between the ladders of rolls while being cooled with water sprays, and then the steel would be bent and withdrawn sideways by a withdrawal machine.

I had brought a camera with me, and I snapped a few pictures of the machine. Some of the German engineers began to look nervous. Franz Küper explained to me in a low voice, "When this machine is set up in Crawfordsville it will be hidden inside a chamber where no one can photograph it." We were looking at the machine in the nude.

Keith Busse climbed a ladder up the machine. His employees followed him, until the machine was festooned with Americans. They clustered at the top of the machine, peering down into the Kowlakowski mold.

The mold was made of copper and formed a sort of mouth at the top of the casting machine. The mouth was a narrow, lens-shaped slit, 4½ feet long and 7 inches wide at its widest point. Into these slightly parted lips liquid steel would be injected in vast quantity, a million tons a year. The inner lips of the mouth gleamed red with the color of copper. A network of high-pressure water pipes was embedded in the copper to cool it; otherwise the copper would melt in a flash when liquid steel touched it.

"This is impressive," said Mark Millett. He stretched out his arms and rubbed his hands together. "It makes you itchy to play with hot metal."

The Germans wanted to show the Americans how easy their machine was to take apart and repair, in the event of *ein Breakout*. The Americans climbed down from the scaffold and stood on the floor, and watched the Germans disassemble their machine. A couple of engineers scrambled up ladders and removed steel pins from the machine. Then an overhead crane picked up the boxlike top of the machine and placed it on the floor. Then the crane picked up two long segments of the machine and deposited them on the floor.

While this operation was taking place, a German engineer said to Keith Busse, "When we install it for you in Crawfordsville, we will test it for

■

you with liquid steel. We will get all the baby diseases out of it for you.''

Busse did not reply.

The engineer went on, insistently, "You must have operators who have lost their fear of liquid steel.''

"We think our people can do a good job with your caster,'' replied Busse.

"We must discuss with you the possibility of bringing in some experienced steel operators from Krupp Steel to run the machine for you,'' said the engineer.

Busse shrugged and said nothing. He didn't want any Krupp steelworkers around Indiana.

There was an argument brewing between the Germans and the Americans. Because Iverson and Marten had worked out the deal so quickly, many aspects of the deal remained unresolved, and had to be covered with riders, subsidiary contracts, and oral agreements. It had begun to dawn on the Germans that the Nucor Corporation intended to hire Indiana farmers to pour hot metal into the machine. Iverson was going to try the same trick that he had attempted when he had started up the Darlington steel mill in 1969. He was going to hire inexperienced country people to run the machines at the Crawfordsville Project. The Germans did not like the idea of Indiana farmers pouring hot metal into their casting machine. In the German opinion, Indiana farmers were not even apprentices, they knew nothing about steel, and they had no business pouring hot metal into an experimental steel-casting machine. The machine could blow up or melt down, that was well within the realm of possibility, and if the machine went up in a mushroom cloud because some idiot of an American farmer pushed the wrong button, SMS would be humiliated before the world steel industry, and SMS would probably never be able to sell another CSP machine to anyone. There were billions of dollars of potential sales at stake for SMS, and it all depended on American farmers, who were supposed to learn quickly how to handle hot metal. The Germans were more than nervous, they were a little bit angry at what they believed to be a certain arrogance on the part of Iverson and the Americans. These Americans were clever with their fingers but they were irresponsible. They were like children. They liked to play with fire, and they had to be watched.

■ ■ ■

Of the Americans in Keith Busse's group, one that the Germans watched closely was a reserved and dignified Nucor foreman by the name of David C. Thompson—Big Dave Thompson. Thompson, a young hot metal man, would be in charge of a group of inexperienced steelworkers hired locally in Crawfordsville, who would pour steel into the SMS casting machine for the first time. Big Dave Thompson had a broad neck, close-cropped ash-blond hair, and calm, serious blue eyes behind glasses, which gave him an intellectual look. He was twenty-nine years old, a year older than Mark Millett. Millett had recently hired Thompson from Rouge Steel, a Big Steel company in Dearborn, Michigan, where Thompson had been a foreman on a classical slab-casting machine. Millett had given Thompson primary responsibility for the first attempts to cast steel. Thompson was the machine's helmsman, the pilot. That was why the Germans were watching Thompson. Big Dave Thompson was six feet four inches tall and he appeared to be several feet wide, but he wasn't fat; he had the marbling of the hot metal man, although the intellectual appearance of his spectacles seemed to contradict the power of his frame. He had been an offensive tackle on his college football team, and he had a master's degree in industrial safety.

Thompson began to poke around the segments of the SMS casting machine with his hands. He inserted a tape measure into the machinery and began to measure the clearances among moving parts.

"I want to learn as much about this machine as I can," he said over his shoulder to a German engineer, who hovered behind him. "I want to put my hands on it." Thompson called for assistance, and two German engineers hurried over to answer Thompson's questions.

"Is this a heat shield?" said Thompson, pointing to a piece of metal.

"*Ja, ja,* of course."

"Good," said Thompson. Thompson wrapped his hand around a water nozzle, which would spray water on the steel. "I like the way these nozzles can be changed," he said.

"*Ja,* that's right," said the German engineer.

"I like that, that's a good system," said Thompson. "What kind of spray water strainers do you have in there?" He reached a hand into the machinery. "It's hard to git at this strainer. I'd like to see something that's quick and easy to pop out. I may not be an engineer, but I have experience around these machines."

"There is the man you must talk to, over there," said the engineer, pointing to another German engineer. "That's Heinz Scholz. He is the chief designer. You must ask Mr. Scholz about the water strainers."

Thompson stood up, hulking large over the Germans. "I'll git him," he said.

■

They backed away. "No! *Don't hit him!* Ask him questions!"

"I'll *get* him, I mean," clarified Big Dave Thompson.

A German remarked to the others that the American had said he was going to fetch the chief designer, not strike the chief designer. Thompson walked across the hall, fetched Heinz Scholz, and Thompson and Scholz returned to the machine.

"This caster has a lot of manual controls on it," said Thompson to Scholz. "In the years to come, can we go fully automatic?"

"*Nee,*" replied Scholz. "We have offered Nucor the more expenzif unit, but you chose the zimple one."

"I've used your competitors' casting machines," said Thompson. "And they're fully automatic. You hit a button and they make steel."

"You can have that from SMS, too! For more money!" said Scholz.

"Maybe the next caster," said Thompson.

"Absolutely! And the next! You can have as many casters as you want!"

Another German engineer took me aside. "These Americans, they are so young," he said in a low voice. "They look like Marines. What is the word—*Ledernecken?*"

"Leathernecks," I said.

"Maybe I have the feeling of the Americans to be a bit . . . what is the word?" The engineer touched his forehead, as if trying to pluck an English word from it. "*Pioneers* is the word," he said. "They are pioneers. You know what I mean? You must look at them. You see? You see these guys with their wagons and their pistols, going to the Wild West." He sighed. "Sometimes I fear the transportation to the U.S.," he said. "We once sent half a steel mill to Australia, and the ship sank!"

I burst out laughing.

He looked horrified. "I am not choking," he said. "The ship sank. This was a disaster for us."

■ ■ ■

At a conference room at the SMS Hilchenbach factory, Keith Busse and his boys sit at a wide table opposite the Germans, for a Negotiation. The Americans wear creamy blended suits and the Germans wear dark, sleek tropical wool suits. In the center of the table, the Stars and Stripes flies beside the black, red, and gold flag of the Federal Republic of Germany, and the flags are flanked by bottles of Cappy soda water, to enable the steelmen to cool their throats in case of shouting.

Franz Küper is reading from a piece of paper, "It is necessary to have tight contact between SMS and Crawfordsville," says Küper. "We recommend a larger number of trained SMS personnel assisting Nucor at the Nucor startup. . . . It means more safety . . . and the shortest period of function testing . . ."

Mark Millett runs a hand through his hair. "It will be *our* operators starting up the caster," says Millett. "It will be *our* people doing the startup, as we've said before."

Big Dave Thompson, the pilot, folds his arms and nods.

Franz Küper listens politely but does not reply.

Keith Busse joins the discussion. "This has been a controversial subject since the beginning, Franz," he says. "Certainly we've never started up one of your casting machines before. And you have. But we are an unusual company. Sometimes when we choose to do things our way, it bites us in the fanny. But it's what sets Nucor apart."

The German side of the table is not pleased by Busse's remark.

Busse continues, "We've never resolved this question of how many SMS people are going to be present at the startup in Crawfordsville, and it could haunt us to the end of the project."

"We certainly agree," says Franz Küper.

Another German adds, "If we don't have enough of our people at Crawfordsville, this could jeopardize the guarantees we are offering—"

"*I object to the reference that this will jeopardize the guarantees,*" says Busse, raising his voice.

The leathernecks nod. Someone passes a pack of Wrigley's Doublemint gum around the American side of the table, and a number of jaws begin to mash chicle. Richard Teets, a bearded Nucor engineer, strikes a match, lights a cigar, and leans back with his hands behind his head and the cigar sticking up in the air.

"Our people are young but they're highly talented," Busse goes on. "They won't let this project slide into the river! It would have been much better if I had some experience in steelmaking. . . ."

The German executives glance at one another.

Busse ignores the Germans. "But I know where Nucor stands. In fact, I was just talking to Dave Aycock, our president, a few days ago, about how many SMS engineers we're going to need during the startup. Dave feels we won't need this many SMS bodies around Crawfordsville."

Evidently the Germans do not like being referred to as "bodies" by an ex-welder and ex-muleteer who happens to be the president of the Nucor

■

Corporation. They consult briefly among themselves. Finally Küper says, "We, er, do have to come to a conclusion on this issue, Keith."

"I agree," says Busse.

Now there is a conference on the American side of the table.

Both sides mutter for a while, until it becomes clear that there will be no agreement concerning the question of how many SMS bodies will be allowed to participate in the startup of the Compact Strip Production machine.

Franz Küper says, "Keith will have to take this back to the, er, States and present it to the Nucor Corporation."

"Let's get on to the hot-strip mill," says Busse, referring to the lower half of the CSP, the part containing the rolling machines, which will be located in the hot mill building under the control of Rodney Mott, the manager of the hot mill.

"We would like to emphasize that we would like to have *two* SMS engineers in the hot-strip mill, but we are allowed only *one* engineer by Nucor," says an SMS executive.

"One man, two men, we're pretty close on the numbers," says Keith Busse, breezily.

The Germans' irritation seems to grow. Franz Küper consults a piece of paper. "We have itemized some other questions of a similar nature here—"

"Now, Franz!" interrupts Busse. "That's just opening an old wound! Now, there again, Nucor's got to *pay* for these SMS salaries. We're not getting your people for free. Why don't you pay these SMS salaries yourself? We're paying you to build us a machine. Then do we have to pay you to make sure it runs right? Time and again I kept hearing from you, 'You don't have to pay for this, you don't have to pay for this'—!"

Franz Küper smiles. "You never heard that from us."

The German side of the table bursts into laughter.

The Americans are not amused.

"Wait a minute, Franz," says Busse. "It's almost like you're assuming Nucor is stupid."

The Germans choose not to reply. There is a clink and a swish as someone pours Cappy soda water into a glass.

"We aren't going to shoot ourselves in the foot," Busse goes on.

"We want you to get a quality strip of steel, Keith," replies Küper, softly.

■

111

"Well," says Busse, "that's why we have all these *Nucor* metallurgists! And all these *Nucor* maintenance people. And all these *Nucor* electrical engineers!"

Rodney Mott, the manager of the hot mill, gazes at the German side of the table with a neutral, unreadable expression on his face. Mott is a soft-spoken manufacturing wizard in his mid-thirties, stocky, dark haired, with rather cool eyes, wearing steel-rimmed glasses, with fourteen years' experience at United States Steel running rolling machines that flatten and elongate pieces of steel. What Rodney Motts wants, most of all, is no Germans bothering him in his mill. "I'd go a step further," he says. "There's people at Nucor who're coming out of the American steel industry who know this stuff as well as you do in Germany."

The Germans are trying to size Mott up; he's quiet and he's hard to read. The Germans are not convinced that Mott is a manufacturing wizard. They do not think that the American steel industry is blessed with talent, considering the obvious fact that the American steel industry recently almost dropped dead. "We are training someone now to help you start the hot mill," says Franz Küper, insistently.

Busse cries, "Who are you training, Franz? Superman?"

"He is somebody to represent our interests."

"Well, if he's going to represent your interests," says Busse, "are you then going to *pay* for him?"

Everyone laughs. Even Küper takes the remark in good humor. "Our man is one who can solve all the questions your experts can't," adds Küper, smiling.

"Well, then, I really think you ought to *pay* for him!" says Busse, also smiling. The smile suddenly leaves his face. He puts his elbows on the table and leans forward. He is wearing a pale-gray raw silk jacket and a white-on-gray silk rep tie. Busse's green eyes seem somewhat paler than his clothes, at the moment. He says firmly, "If you're not going to let me cut down on the number of SMS bodies in Crawfordsville, how about letting me cut down on the total amount of *time* they spend in Crawfordsville?"

Annoyed silence radiates like heat from the Germans in their dark suits. It is evident that the Germans still do not accept being referred to as "bodies." After a lengthy pause, Franz Küper diplomatically replies, "*Ja,* we can consider the budget of Nucor." He looks around his side of the table and receives nods of affirmation from the other Germans. "Possibly we can agree to reduce the number of man-hours our people spend in Crawfordsville," he says.

"Fine, we'll agree to that!" says Busse. "Are your people going to work *two* shifts a day? We're working sixteen hours a day out there in Indiana to build this mill. We're running two shifts a day. Are your people going to run two shifts right alongside us? There's no such thing as a forty-hour workweek at Nucor. Your engineers are going to have to work a forty-eight-hour week, minimum, and in two shifts per day."

"This is another thing, Keith," replies Franz Küper, in a cool tone of voice. "You now want a forty-eight-hour week from our people. You never spoke of this before. We are calculating all this according to a forty-hour workweek."

"You mean, Franz," says Busse, "you pay your people extra to work a forty-*eight*-hour week? Good heavens, we don't do that at Nucor!"

Smiles ripple around the American side of the table.

"Our people have to be paid," says Küper. "And we are going to pay them overtime."

"You mean, *we* are going to pay them overtime," says Busse.

"Our people have to be paid," repeats Küper. "And they will be paid overtime, Keith. Should we, er, put this in writing?"

Busse pauses for a moment. He appears to be wondering whether to fold his cards or to double the stakes. Busse says, "So what was the agreed-upon billing rate?"

"Eight hundred dollars a day per SMS engineer, including airfare and expenses."

The Americans, shocked or feigning shock, sit back from the table and stare. "My goodness, Franz!" says Busse. "We're talking about a large amount of money here! Eight hundred bucks a day? Run that through your calculator, Franz. You could be a million dollars over our budget."

"We don't make money on this," replies Küper. "We make money selling machinery."

"Yeah," says Busse, "but I would say that eight hundred bucks a day per SMS person is a pretty stiff rate."

Mark Millett runs a hand through his hair and breaks in: "Compared to our other vendors, it's phenomenal! We've got others working for us for four hundred dollars a day."

"We can put your people up for twenty-five bucks a day in our famous

■

Crawfordsville motel, the General Lew Wallace," adds Busse. "The General Lew is a nice motel. It has a bar. Your people will like it. Maybe you should reduce the rate you are charging—eight hundred dollars a day seems excessive. If you charged less for your time, that would be a way of getting more of your people to Crawfordsville."

The German executives ponder Busse's idea for a moment, until one of them remarks, "So why don't we prepare a 'lump sum' proposal to Nucor? Instead of a daily rate for our engineers." The Germans lean across one another and huddle for a moment, speaking in German. The words "*Die Lumpsum, die Lumpsum,*" drift faintly to the American side of the table, and then "*Nee! Nee!*"—the Germans have decided *not* to offer their engineering talent to the Americans for a lump sum. The Germans want their engineers to be paid by the day, and they want as many SMS engineers as possible at Crawfordsville to prevent Indiana locals from immolating their beautiful machine in a hecatomb of boiling metal before the eyes of the world steel industry. Franz Küper knocks a stack of papers on the table to align the papers and says, "Shall we, er, go on to point two, then . . . ?"

■ ■ ■

The Negotiation ended in a bowling alley. Bowling alleys in Germany serve food and hard liquor. The final point of the Negotiation was to determine which nationality could drink the largest amount of liquor—the final and most important point of all Negotiations among steelmakers. The bowling alley in Hilchenbach is a converted railway station. A little train still makes a whistle-stop there twice a day. There are several alleys inside the building, separated by movable partitions. The partitions form private dining rooms, and the rooms look out on the alleys. The Germans had rented a private dining room for the affair. They sat at one table, along with Keith Busse, and the rest of the Nucor people sat at another table. Waitresses carried in bowls of brown, spicy soup, setting them down on the tables over the sounds of bowling pins being knocked around.

"What is this?" says a Nucor guy.

"It's oxtail soup."

"It looks like they boiled the hell out of it, anyway."

"I'm lookin' for a tail, here."

"It's not the tail I'm worried about. I'm worried about somethin' that could come attached to the tail."

"Oh, God, shut up!"

■

The waitresses brought in a round of the local beer in mugs. Paper coasters under the mugs informed us that the beer had been brewed from the Famous Felsquellwaters. The Famous Felsquellwaters drowned the oxtail soup. Another round of beer arrived, and another, and next arrived plates of *Schnitzel*—slabs of deep-fried veal.

Franz Küper stood up at the German table. "Your attention, please," he said. "In this first game we'll be playing for the lowest number. The object is to knock down as few pins as possible. But you can't roll the ball into the gutter."

The Americans and the Germans formed national teams, and the Germans destroyed the Americans at low-ball bowling.

"At least we made second place," said one of the Nucor men.

"You have lost," loudly declared a German engineer by the name of Dieter Richstein. "Now you must pay us one bottle of whiskey extra per month, per engineer. That will be in our contract."

Busse said in a low voice to the Americans, "How do we even this up in ten minutes?"

Another round of beer arrived.

"For this next game," said Franz Küper, "the idea is to hit the numbered pins in sequence, everybody."

The Americans couldn't hit anything in sequence except the gutters, and they lost the game dramatically.

"Now for the next game," said Küper, "the even-numbered pins count single and the odd-numbered pins count double. It's, er, more complicated."

It certainly was, especially after another round of beer had gone down. The Germans diplomatically suggested mixed German-American teams.

Dieter Richstein, the engineer who had earlier claimed whiskey in his contract, raised his beer glass for a toast. "I have announcement to make," he said, getting to his feet. "My first impression . . . my first impression of our working relationship with these American steelmakers is that . . . is that . . . *they are strong drinkers.*"

We drained our beers to impress Richstein. Now Richstein crossed the room and sat down at our table, bringing with him a tray of little glasses filled with clear spirits. "This is schnapps," said Richstein. "You must drink it."

We drank it.

■

Richstein waved his hand. Immediately a waitress appeared with another round of spirits.

"What's this?"

"*Noch ein Schnapps,*" he said.

"What'sat mean?"

"That means, 'Another schnapps,' " said Richstein.

We drank it.

Richstein peered into our faces with deep interest. Richstein was a small man, an expert on rolling mills, with a beak nose and hair going white. He called for a little something extra. When the little something extra had arrived at the table—another clear liquid in little glasses—he said, "We are having *Pflaumenschnapps*. You must drink it."

"Flawm what?" said a Nucor guy.

"Plum schnapps! You must drink it."

We drank the plum schnapps.

Richstein called for a round of something new, and while he waited for it to appear, he leaned across the table and gave us a lecture on our responsibilities as Americans. "Before the War," he said, "all the new ideas for steel plants came from the United States. After the War, there has been nothing from you. Nothing! Now all the papers on steel are written by Europeans and Japanese. So what has happened to you?" Richstein shrugged and gave us a look full of meaning, as if to say that the American failure in steel was one of those ugly secrets best not examined too closely by America's friends, just as you would not ask a business associate who was rumored to have a brain tumor to tell you the results of his latest CAT scan. "So now it is time for America to be saved by Americans," he said.

Nothing could save any Americans at the bowling alley. I bowled four gutter balls in a row, crippling a Nucor team. I wound up to throw the ball again, and Rodney Mott, the hot mill manager, said to me, "Come on— throw it nice and easy, now! At Nucor we're not used to coming in last." So I threw a nice-and-easy gutter ball. Jeers and catcalls from the steelmakers—"Five gutter balls in a row! We're gonna puke the reporter!"

A waitress brought another round of schnapps to the table. "This is *Himbeergeist,*" Richstein declared. He told us that it meant "raspberry ghost."

The raspberry ghost was a clear white liquor that smelled invitingly of summer berries. We swallowed it. The stuff was hard as quartz.

■

A thick, oozy schnapps that stank of barley appeared.

"This is *Kopfschmerzen,*" declared Richstein. "That means, 'Headacher.' You must drink it."

We tossed it off, but did not notice any immediate effects.

Shot glasses appeared, marked with sinister black crosses. "It's good for you!" cried Richstein. "It's Maltese Cross!"

We drank the Maltese Cross. The stuff tasted of anise.

Richstein spoke to a waitress, and then said, "Now we are having *Korn.*"

"Corn, what's that?"

"*Korn.*"

"Corn liquor?"

"*Ja!*"

"You mean moonshine?"

"*Ja!* There you go! Moonshine," said Richstein.

The American steelmakers looked at each other in surprise. They hadn't suspected that German steelmakers like to drink moonshine, too.

A collection of little glasses appeared, containing *Hilchenbach Korn,* a raw, white liquor fresh out of a local pot still. We drank the *Korn.* The *Korn* immolated our soft palates. They burst into flames like oil-soaked rags, and meanwhile, Richstein studied our faces with the intensity of a forensic pathologist. Someone cried out for beer.

The German engineer hurried away to get a round of beer, to help us douse the wildfires raging in our throats. In a few minutes he returned with a tray of beers, and we drank them off quickly, whereupon Richstein declared that the beer had been mixed with schnapps. "Do you like it?" he said. "The other beers you have drunken tonight had schnapps, too."

A dull shock went around the American table, as it dawned on the Americans that they had been drinking German boilermakers and hadn't realized it. This was supposed to be a drinking contest between the vendor (SMS) and the customer (Nucor), but the question seemed to be whether the customer could drink itself under the table.

At the beleaguered Nucor table, Mark Millett muttered in a thick British voice, "I was pissed at Franz Küper! What he said about us needing so many SMS people on the caster! Tomorrow I'll wake up and say, 'Yeah, it was a dream.' " Millett slapped his own face, as if to wake himself from a dream.

Meanwhile there was trouble brewing between Big Dave Thompson, the machine's pilot, and a certain Kristy E. McGee, the manager of construction of the Crawfordsville Project. Kris McGee, thirty-two years old, had a ruddy face, a wire-brush mustache, and a wonderfully loud voice. He was a refugee from U.S. Steel, and he was reputed to be one of the more brilliant civil engineers in the American steel industry. McGee's specialty was building factories.

Big Dave Thompson, a foreman and hiree from Rouge Steel, was not unaware of the fact that if Nucor's casting machine melted down, the eyes of the world steel industry would focus on Dave Thompson, who had poured metal into the machine. Both McGee and Thompson had come out of Big Steel, out of rival companies.

People at the table began to tease Thompson. Someone said to Thompson, "Half the steelworkers at Rouge Steel are on crack and the other half are so dumb they don't know their own names." The remark did not sit well with Big Dave Thompson. The discussion turned to business degrees. Mark Millett asked his employee, Thompson, if Thompson happened to have an M.B.A.

Thompson answered in a dignified way that he had a graduate degree in industrial safety management.

"Safety management!" roared Kris McGee, the construction manager. "A master's in safety!" That was great, he said, because a master's degree in safety was just what Big Dave would need if the caster blew up in his face.

Thompson looked like he was about to blow up. He stared at McGee, and then stood up. He raised his beer mug and said, "To massive production!"

The Americans stood up and drank to that sentiment, and then an idea solidified among them that they should return to their hotel for a drink, the American steelmakers being of the opinion that bowling alleys in Germany don't offer enough to drink. They crowded into rental cars, accompanied by their sober and nervous handler, Franz Küper, and they roared down the main street of Hilchenbach, roared up a hill, parked at their hotel—the Hotel Sonnenhang ("Hotel Sunnyside")—and jammed into the bar.

The Sunnyside was a white hotel on a mountain ridge, surrounded by rhododendrons, where elderly German tourists like to take their vacations. The bar was a small room where the elderly German tourists liked to drink *digestif* and speak of the beauties of the mountains, but when the Americans arrived, the room went opaque with bulky cream-colored suits and

■

began to rattle with deafening shouts. Keith Busse declared that he was going to bed.

As soon as Busse left the room, the Americans ordered a round of *Korn*. The Sunnyside's bartender dosed out *Korn* in little fat glasses, and the glasses were hoisted in the air.

"To steel by next spring!"

"To a long slab, no breakouts!"

Kris McGee said something about safety management on a casting machine and suddenly McGee was hoisted in the air. Thompson picked up McGee by the shirt and threw him against the bar with a ripping sound. Thompson said, "Do you want to settle it outside?"

"There's nothing to settle," answered McGee.

The crowd froze, like one of those crowds described by Tacitus, transfixed at the sight of a general about to be murdered by one of his own troops, until suddenly the Americans converged into a struggling knot along the bar. Franz Küper backed into a corner of the room and the bartender faded away to call the police.

Mark Millett came down between the two steelmen, shouting, "Hey! Hey! Hey!"—pulling McGee away. Others stood around Thompson.

There was a moment of stillness. Thompson offered his hand to McGee. "I apologize," said Thompson.

"I don't need to apologize," said the manager of construction. "I didn't do anything to you. There's nothing to apologize for."

"That's bad form, Kris," said Mark Millett. Millett took McGee aside. "What the hell did you do to him, Kris?" said Millett.

"You guys were bugging him, too," said McGee.

The crowd broke up, and the steelmakers lurched upstairs to bed. McGee was dreadfully embarrassed. He felt that he had acted badly with Thompson. As for myself, I became afraid of something that night. The steelmakers were nervous. They had been given a chance to build a steel mill that was supposed to change the world steel industry, and the Project that Iverson had put in motion was beginning to sweep them along, beyond anyone's power to control it. I began to feel that the price of making steel was going to turn out to be higher than anyone expected. When someone dies from contact with steel, you think afterward that you saw how it almost had to happen, but then maybe you didn't see it coming, maybe you were just afraid of hot metal.

■

9

A PRACTICAL
DEMONSTRATION

■ ■ ■

The next morning the elderly German tourists at the Hotel Sunnyside were awakened by racking coughs coming from the wing of American rooms. Toilets began to gargle and that continued for a long time, mixed with sporadic deep-drawn gasps and muffled curses. There followed the quiet of the morgue, until a few steelmen drifted into the dining room and slumped at tables, staring at cups of coffee. Someone remarked in a feeble voice that Nucor was selling Buicks in Germany this morning— "*Buick! Buick!*"—the sound of a steelmaker clearing his throat before breakfast.

Keith Busse had received an exemption from agony. Apparently immune to *Korn* and boilermakers, Busse grumbled about the near-brawl in the hotel bar last night. "That was inexcusable," he said, pressing a fork into a heap of scrambled eggs. "But think what it's like for these guys. This company's putting them under unbelievable pressure," he said, chewing egg. "I don't know that we can even imagine the pressure these guys are under. Plus they had a bit too much to drink." Watching him chew egg, I felt a certain pressure to sell a Buick myself.

We were supposed to watch Manfred Kolakowski give *eine praktische Demonstration,* a practical demonstration, of the art of casting steel into a thin slab. The Americans drove in a tight string of cars at high speed through the town of Kreuztal and turned left into the village of Buschhütten. Buschhütten means "foundry in the woods." The Americans drove down a boulevard shaded with beautiful old plane trees, and suddenly they

■

arrived at a cobbled parking lot beside a collection of old, gray buildings, under a gray sky. This was the foundry where Kolakowski worked his wizardry. Swallows concirculated in the air around the building. We went into a conference room to wait for Manfred Kolakowski to appear. The floor of the conference room rumbled.

"Listen, you can hear it," said Keith Busse.

Kolakowski was melting steel.

The Americans drizzled coffee down their throats and spoke in thick voices:

"When I woke up this morning, my head was bigger than a blue-ribbon hog."

"This is it, the making of steel."

Manfred Kolakowski entered the room and shook hands all around. Kolakowski's title at SMS was Senior Group Leader for Technology. The inventor was in his late forties. He had a long, massive, severe face with a high, prominent forehead, bags under his eyes, and a gravid jaw that hung forward and down. He fixed a deep gaze on Keith Busse. "How is it going, your work in Crawfordsville?"

Busse replied that the land had turned into a sea of mud during the past winter, and that consequently the job was behind schedule.

Kolakowski nodded.

"How's everything going here?" Mark Millett asked Kolakowski.

"Oh, *ja,* not so bad," said Kolakowski. "Today there is a little bit of nickel in the steel. Sometimes it works, sometimes not. You want to make progress, so you try different things."

"How many breakouts have you had, so far?" asked Busse.

"I think about forty breakouts in total, out of four hundred casts," answered Kolakowski, who spoke fluent English. He led us across a cobbled parking lot to the main foundry building. Inside the foundry, in a relatively small space, stood a casting tower three stories tall. This was the prototype, Kolakowski's plaything, the pilot casting machine. The machine itself was concealed inside the casting tower. A roar and a flicker of blue light coming around a corner in the foundry announced the presence of a small electric arc furnace, which at the moment was finishing a heat of steel—making a bath of liquid steel.

Kolakowski and the Americans put on aluminized suits called "silvers," designed to protect them from splashes of hot metal. They put on silver coats, silver gloves, and silver booties, and they put on hard hats with clear faceshields. Suited up like astronauts, they enjoyed slight pro-

■

tection from a splash of molten steel. But steel splashes may be tremendous and may be accompanied by other phenomena that a silver suit may or may not provide any protection against, such as high-voltage electrical fires, superheated steam explosions, flesh-cutting sprays of hydraulic fluid, concrete explosions (concrete explodes if liquid steel is dumped on it), and chunks of machinery the size of engine blocks pinwheeling through the walls of a plant. The casting temperature of liquid steel is 2,850 degrees Fahrenheit, about one-quarter of the surface temperature of the sun. Kolakowski spoke of the difficulty of controlling the experimental machine when there was liquid metal inside it. "All decisions have to be made in seconds," he said.

The top of the machine was surmounted by a platform with a railing around it. This was the casting deck. The casting deck sat across the top of the casting tower like the bridge of a ship. The deck was made of concrete, and it had escape stairways at three corners in case the ship had to be abandoned. The center of the deck was cut by a scorched and blackened slit. This was the mouth of the Kolakowski mold. The mouth was a funnel lined with copper. The rest of the machine extended below the casting deck, beneath the feet of the steelworkers. The casting deck is the most dangerous place in a steel mill, because you are standing on top of a large machine and below a full ladle of steel, which is pouring a river of steel into the machine.

A sort of bathtub called a tundish squatted on legs on the casting deck, above the slit. The tundish, the bathtub, was an intermediate holding tank for liquid steel. Steel drained in two steps into the machine. First it dropped from a hole in the bottom of the ladle into the bathtub. (A ladle is a giant bucket for carrying liquid steel.) Then it ran down into the machine. So the bathtub filled up and then the steel drained from the bathtub into the machine. A white ceramic tube, an injection nozzle, extended downward from the belly of the bathtub and deep between the lips of the Kolakowski mold. The nozzle fed hot metal into the mold, the mouth of the machine. The mold was red on the inside and clearly female in structure. The injection nozzle was creamy white in color, made of a ceramic material, and it was long, slender, fragile, and obviously male, and it fitted neatly into the Kolakowski mold.

The arrangement was copulatory. As liquid steel flowed through the injection nozzle into the machine, the Kolakowski mold would start to move up and down, urgently, faster and faster, as the cast was accelerated and steel moved through the machine. This sort of oscillation is typical of

■

casting machines. It prevents liquid steel from sticking to the inner faces of the mold. "If you see the straight tube going into the vagina-shaped mold, you know that many jokes can be made about this," Kolakowski said.

The Kolakowski mold marked the dividing line between liquid and solid metal, the critical point of steel; it was the womb of a steel mill, the organ of procreation. Below the mold, a solidifying thin slab of steel would travel vertically downward between the machine's rolls, and then the slab would be bent and withdrawn horizontally from the base of the machine.

Until Kolakowski dreamed up a mold shaped like a letter envelope that has been squeezed open, all molds in classical steel-casting machines had been boxlike funnels with flat sides. The experts had told Kolakowski that a swollen curved mold, with a queer shape, a biological shape, would certainly extrude a cracked and warped piece of steel. The experts had been wrong. Kolakowski's lens-shaped mold produced smooth, flat, thin, rectangular steel.

I asked Kolakowski how he had got the idea for the mold.

The inventor could not quite explain how. "It happened to me very suddenly," he said. "It happened on the eighteenth of March, 1983, and I know that it was the eighteenth of March, because it was one day before my Easter holiday." Kolakowski had seen the mold in his mind's eye. He had sketched it on a piece of paper. "Then I went with my children to Milan, to see the spring."

A bright glow appeared beside the casting deck, as a crane brought over a ladle full of live steel, restless liquid metal, horripilating with dissolved oxygen, a heat of steel. Manfred Kolakowski lowered his faceshield and began to give orders in German to his men, and the casting deck rumbled: high-pressure water pumps had begun to circulate coolant water through pipes embedded in the Kolakowski mold.

The ladle, the bucket for carrying steel, had the shape of a Dixie cup. A white blanket of limestone slag floated on the steel in the ladle, like a head on a beer in a Dixie cup. Occasionally the blanket cracked to reveal naked metal, and through the cracks an orange-white light burst and played across the ceiling of the foundry, bathing our bodies in warmth. You could not look directly at the naked metal without pain in the eyes. A hissing, crackling cough came from the ladle. It was the deep-fat snarl of a Pitco Fryalator cooking on high. At the same time a metallic timbre of iron reverberated deep in the ladle, the unsatisfied mutter of live steel wanting to go into a carbon boil. The sound of the metal did not inspire confidence in the gadgets of inventors.

■

German steelworkers, standing on a platform, shoveled silicon sand into the ladle to "kill" the steel. The sand removed the dissolved oxygen from the metal. Liquid steel that contains dissolved oxygen is live steel. Live steel can foam out of a ladle, burning people to death. Live steel cannot be poured into a casting machine, because it will foam inside the machine. Only killed steel can be poured into a machine.

The Americans, wearing silvers, stood to one side on the deck and lowered their faceshields. Big Dave Thompson stood comfortably at ease, watching the hot metal with a calm gaze through his spectacles, and the light of the metal shimmered on Thompson's clear faceshield. No one had offered me silvers to wear, but I asked Franz Küper if I could stay on the casting deck with the Nucor steelworkers when the cast was performed.

He shook his head. "These Nucor guys know liquid steel, and you don't," he said. "If there's an emergency up here, they know what to do. We do have emergencies."

Küper led me off the casting deck, to where Keith Busse stood by a railing. None of us were hot metal men, and so we had to watch the play from a safe distance.

The crane picked up the ladle and moved it over the casting deck. The ladle hovered over the tundish, or bathtub, above the mouth of the machine. There was a small black pipe sticking out of the circular bottom of the ladle. That was the hole from which steel would pour into the bathtub. A German steelworker fitted a sleeve over the black pipe, and someone shouted an order. A flash of light appeared at the pipe, and a stream of steel ran through the sleeve and into the bathtub, the tundish, with a roaring sound and a rich burst of light and sparks.

"Now the tundish is filling with steel," explained Küper.

Kolakowski, pressing buttons on a computer control panel, opened a stopper-valve in the bathtub. That allowed steel to drain down into the casting machine through the male injection nozzle. There was another burst of sparks.

"Now the mold is filling," said Küper.

There was a pause. A shower of sparks flew into the air over the casting deck and fell among the steelworkers.

"Now the cast has begun," said Küper.

A red thin slab of steel began to emerge like a tongue from the base of the machine.

Suddenly flames shot up from the casting deck. A cauliflower of smoke boiled to the roof of the foundry. German steelworkers began to run in all directions across the deck, shouting in German. The machine appeared to have caught fire.

"I think it is an emergency," said Küper, in a flat voice.

A volcanic eruption started from the mouth of the machine. Smoke began to pour from the sides of the machine.

"Something's *fucked*," said Busse.

A siren went off, *whoop, whoop, whoop!* The crane suddenly lifted the ladle up and away from the casting deck, to get the ladle away from the casting tower, in order to stop the flow of steel into the machine. This is known as an abort procedure.

Germans in silvers ran back and forth across the casting deck, shouting, and the Americans gathered near the exit stairs, getting ready to run. A skunky odor hung in the air, the reek of burning steel, as the steel inside the machine caught fire and burned. The injection nozzle had snapped off and fallen down inside the machine—a nasty sexual-industrial accident. It was a breakout, a rain of hot metal down through the machine.

"They've aborted the cast," said Busse. "It's all shot to pieces."

That made it forty-one breakouts for Manfred Kolakowski, and the siren going *whoop, whoop, whoop* as it laughed at Kolakowski's attempt to satisfy the hope of Sir Henry Bessemer, who first imagined the Machine during the reign of Queen Victoria.

10

THE FAST TRACK

■ ▪ ■

The Compact Strip Production machine was packed into more than four hundred wooden crates during the summer of 1988, in Hilchenbach, West Germany. Many of the crates were larger than house trailers. The crates were loaded onto special railroad cars, which carried the CSP to the port of Bremerhaven on the North Sea, where the CSP was loaded into ten river barges. Then the barges were lifted whole out of the water by shipping cranes and were loaded into two oceangoing freighters. The plan was for the freighters to cross the Atlantic Ocean and dock in New Orleans, where the barges would be placed in the Mississippi River and pushed by tugboats up the Mississippi to the Ohio River, then to southern Indiana.

A drought across North America that summer lowered the Mississippi River and grounded barge traffic on the river. The two German freighters altered their course and passed through the Saint Lawrence Seaway into the Great Lakes. They docked at Burns Harbor, Indiana, on the southern shore of Lake Michigan, in the middle of the Lake Shore steel mills. Nucor's CSP was unloaded in the solar plexus of Big Steel.

The Press Express, a trucking firm out of Chicago that specialized in the transport of capital machines, delivered the CSP's roll housings to Crawfordsville. The roll housings were the heaviest parts of the CSP, the solid-steel doughnut rings for the millstands—26 feet long, 120 tons apiece. The Press Express trucks had nineteen axles, seventy-four wheels, and two cabs, one in front and one in the rear. The driver in the rear cab steered the truck's rear wheels by means of a hydraulic ram. The Press Express trucks

■

made a winding progress toward the center of Indiana, avoiding weak bridges. The first roll housing arrived at Crawfordsville on October 6, 1988, one year after ground had first been broken for the steel mill.

The job site became a sea of machinery. Delicate pieces were stored inside the melt shop and the hot mill, and heavy goods were placed outdoors in a staging area to the south of the melt shop. The Crawfordsville Project began to resemble the staging grounds in southern England in the months preceding the D-day invasion of Europe, in the spring of 1944. A feeling came over the Nucor steelworkers that they were engaged in an industrial war.

■ ■ ■

Keith Busse was under immense career pressure. There was no question but that Keith Busse was being trained for possible leadership of the Nucor Corporation, but Busse wasn't the only candidate for the job, not by a long shot. Nucor was also building a steel mill in Blytheville, Arkansas, a joint venture with a small Japanese steel company, Yamato Kogyo Co., to produce I-beams for skyscrapers, made from melted automobiles. Even while Iverson was talking about a leapfrog of Japan, he was doing deals with Japanese steel companies.

"Our Japanese partners love us," claimed Iverson.

He figured that the Blytheville Project would ram so many I-beams down the throat of the American skyscraper market that the price of I-beams would crash, and Nucor would thereby drive at least one Big Steel company out of the I-beam business.

The Blytheville Project melted its first heat of steel on July 4, 1988. The furnaces took a nightmarish fifty-six hours to fire up, but soon afterward, the mill was manufacturing raw semifinished blanks for I-beams. By autumn, the mill's rolling machines had started up, and the mill had an order book full of orders for I-beams. The Blytheville Project was built and started up by a Nucor general manager named John D. Correnti. Correnti and Busse were watching each other like hawks, and Correnti was having big success down in Arkansas. Their last names rhymed—Correnti and Busse—and so did the names of their respective Projects—the Blytheville Project and the Crawfordsville Project.

"I hate the heir-apparent syndrome," said Iverson, and he made it clear that any manager whom he caught jockeying to become CEO of Nucor would get into trouble with F. Kenneth Iverson. But Keith Busse and John Correnti, being middle-level managers with large ambitions, were in strong

■

competition with each other to prove to Iverson and to the company that they could build a steel mill of the future.

As soon as the Blytheville Project went through startup, all eyes at the Nucor Corporation turned toward Keith Earl Busse and the Crawfordsville Project. The Crawfordsville Project, in the fall of 1988, was supposed to be less than eight months from startup. Yet the job site was a horror, the machines were sitting around in pieces. Busse felt pressure mounting on his career, on his family life, and on himself. The Crawfordsville Project marked Montgomery County that summer with a plume of dust, thrown up by bulldozers and cranes moving back and forth across a desiccated mudscape as hard as porcelain. Duane Hurler hung his iron and went home to his honey and his coon dogs, but a Red River crew remained at the job site to cover the melt shop with Nucor prefabricated steel panels. Summer waned in Crawfordsville, hummingbirds abandoned sycamore trees along Sugar Creek and flew away to Mexico, and the sycamores dropped brown leaves as big as shopping bags into the stream. With the arrival of cool weather and shorter days, a galaxy of sodium lights went on all over the Crawfordsville Project, and the work went on deep into the night.

■ ■ ■

On a calm day in September, a Red River iron hanger by the name of John Young was working on a crane rail in the melt shop, eighty-five feet above the ground. A crane rail is a girder, eighteen inches wide, upon which a traveling crane moves. John Young was using a pry bar to try to loosen a piece of structural steel. The pry bar slipped and he stepped backward into space. He had not clipped his safety strap to anything. People heard him shout. He fell eight stories inside the melt shop, looking up at the roof as he fell. He landed on his back.

Will W. Hawley, the utility man who had accompanied me on the crane drop, arrived at the scene of the iron hanger's fall. The result of an eight-story drop on the human body is not a pretty thing to see. John Young's eyes were still open and he was looking up at the roof and at fair-weather clouds running across an open bay, and his head was soft and misshapen. "His neck was broken and blood was coming out of his eyes," Hawley later recalled. Blood poured from the iron hanger's open eyes like tears and ran down his cheeks. The iron hanger seemed to be grieving. Then the tears stopped flowing, because his heart had stopped. *When you don't get it between your ears that you can be killed up there.* John Young was buried in Texas at the age of nineteen. He had a wife and a baby. It was the first death at the Crawfordsville Project.

October deepened in Indiana, and the maple trees ignited and burned and went bare. Cold rains pelted the porcelain ground, and it became bottomless clay. At the western edge of the Nucor property, goldenrod withered and the margin of the woods filled with dry racemes, umbels, burrs, and stickers; and sylvan pear trees offered mottled fruit. Nucor men drove pickup trucks out to the woods during their lunch breaks and gathered the pears. The pears were unaccountably sweet. Turkey vultures sailed over the mill and collected by the dozens in thermals over stubbled fields, and traced helixes within the thermals as they ascended toward the clouds, holding their wings in a dihedral V, with their tail feathers tucked together, which produced a rocking motion in their bodies as they climbed.

■ ■ ■

Big Dave Thompson, foreman and pilot for the casting machine, wandered among machinery in the staging area to the south of the melt shop, with a toothpick in his teeth and a hard hat on his head. It was an October afternoon, and ditches around the machinery were full of gray water that reflected a Midwestern autumn sky the color of plain carbon steel. A smell of decayed leaves hung in a windless air. Thompson carried a bundle of blueprints in his hand. He unrolled it and stared at a print.

"There are about thirty-five major components to the system that makes up the casting machine," he said. "They came in about a hundred and fifty boxes. All the parts of the caster are here at Crawfordsville, somewhere. What I got to do right now is *find* 'em. It's already taken us a week just to locate the motors for the casting machine. We put the motors in the melt shop. They'd have been ruined if they were left out in the rain."

He stared at a print. "What are the drawing numbers on these prints?" he wondered, flipping through the prints. "Here we are," he said, reading aloud some numbers. Then his eyes traveled over the machinery and stopped on a heap of rusty pipes.

"O.K., this pipe came from the old Klöckner mill in Germany. That'll go in the cold mill." Thompson squatted down to examine a collection of steel frames. "O.K., these are machinery mounts." He switched the toothpick around in his teeth. Thompson had a massive, wide, almost cylindrical back. He straightened up and walked around behind the machinery mounts, where he discovered nine bathtubs, chest high, sitting on the ground.

■

"These are the tundishes," he said. One day next spring, Thompson and his fellow steelworkers would fill a bathtub full of steel and let the steel drain into the machine.

I asked him if he had had much experience in the steel industry, and he gave me a funny look. "I grew up in East Liverpool, Ohio, on a farm just above the River," he said. He meant the Ohio River, which ran through the broken heart of steel country. "We had a family farm in rolling hills. My grandfather farmed dairy cattle on it and wheat and grains." Thompson's father and uncle inherited the farm, but they both worked in steel mills while continuing to run the farm. "My father worked for thirty years at the Crucible Steel Company, on the other side of the River. He was a melter. I spent my summers in there, making money to go to college. That's where I got fascinated. I never thought I'd have a chance, after what happened. I'm just glad to have a chance." Thompson had grown up on a dairy farm, but he was a hot metal man of the second generation.

A pickup truck screeched to a halt near us. Mark Millett leaned his head out the window. "Wot's happening?" he said.

"Taking inventory."

"So have you put in the casting machine yet?"

"Not yet. We're working on it."

"Chaos, I love it," said Millett, and his truck shuddered away.

■ ■ ■

One evening, Busse parked his car—a white Audi with smoked windows— in front of the L&M Lounge in Ladoga, where retired farmers hang out along a bar, eating hamburgers and drinking a little whiskey. At a table in the back of the lounge, Busse found a group of Nucor managers, including Mark Millett (melt shop manager, upper part of the CSP) and Rodney Mott (hot mill manager, lower part of the CSP). They were eating hamburgers and drinking beer. At the center of the group sat Dieter Richstein. He was the same Dieter Richstein who had force-fed the managers moonshine at the bowling alley in Germany last June, but now, in October, he was not the same Dieter Richstein. He was subdued, unhappy. He was a person of a slight build, in his mid-fifties, with whitening hair and a birdlike nose, and he wore a poplin sporting jacket.

Richstein was saying in a low voice to Busse's managers: "We have been making too many compromises in the engineering, do you know that?"

Nobody replied.

■

Busse ordered a beer and a hamburger.

"Compromises, compromises, compromises, *ja*?" said Richstein, sipping a beer.

Busse looked annoyed. "Dieter!" he said. "When you first came over here, you said we'd be making steel by next spring. What do you think now?"

"*Compromises,*" said Richstein, darkly.

Busse glared at him. "Dieter, let me ask you a question. You are talking tonight in a very pessimistic vein. But whenever I see you in the construction trailers, you sound optimistic. How come? How can you be optimistic in the trailers and pessimistic at this dinner table tonight?"

"I am dismayed," said Richstein. "We have had good luck with the weather so far. But when the rain comes this winter . . . We think about the rain. We think about settling of the machines. We think about how we can set the tolerances on the machines with a millimeter screw when it's raining."

"Like I've been saying, screw it down and pull steel through it!" snapped Busse.

The managers laughed.

Richstein only smiled in a troubled way. "When the circumstances are so bad," murmured Richstein.

"Dieter, the circumstances are what they are," said Busse. "This is Nucor. *That's a circumstance right there.*"

Richstein bowed his head and pursed his lips.

Busse went on: "This is *Nucor,* Dieter. This is Nucor wanting to finish a steel mill one and a half years after breaking ground. That's the Nucor way and you have to get used to it."

Richstein brushed his hands back and forth across the table, dusting crumbs from the table. "*Ja, ja,*" he said, "but you have to realize that the Nucor way is not the *normal* way." After a pause and more dusting of the table, he added, "I am worried about all the dirt. The dirt is everywhere. The dirt will get into the rolling machines."

"You are going to have dirt no matter what," said Busse. "In the melt shop, the floor is going to be *made* of dirt. You know that! Nucor doesn't use concrete floors in its melt shops. You know that! Nucor is running melt shops all over this nation with dirt floors."

"*Ja,* okay, in the melt shop!" Richstein's voice rose almost to a cry of pain. "But the dirt in the *hot mill,* the dirt in the *cold mill*—"

"All right, and I'm telling you, it's a sickening mess! So what? Some-

■

times I think I just want to pour those floors and then cut holes in them with a jackhammer, for the plumbing!'' Busse's beer arrived, followed by his hamburger, and Busse took a pull of beer and chomped into the hamburger.

Meanwhile Richstein looked nervous, as if he believed Busse might just pour the floors and then cut holes in them with jackhammers afterward, and then insert the plumbing into the holes. Yes. That was exactly the kind of thing a Nucor general manager would do.

Mark Millett tried to calm the waters. He said, quietly, "Dieter, the thing is, if you want to get the best prices for your equipment, you have to let companies like Nucor be creative.''

Richstein said nothing.

Busse swigged his beer and added, "I understand that the way we do things is not your way. I understand that.''

No, Keith Busse did not understand the German way at all, in Richstein's opinion.

Busse said, "We've had to do the engineering for this mill as we walked along. That's the way we do a mill, Dieter. It is unacceptable to this company to develop beautiful engineering plans, and to pour beautiful concrete floors, and then to paint the floors, and then to install equipment on the painted floors. That method is *unacceptable* to Nucor.''

"I don't think our quality has been compromised, Dieter,'' said Millett. "We can still make an excellent product. We will make steel that anyone could be proud of.''

Busse turned to Rodney Mott, the manager of the hot mill. "When *are* we going to finish pouring the God damned floor in the hot mill?''

Mott didn't exactly answer the question. "Just keep on pouring concrete until the floors are finished,'' Mott said.

Dieter Richstein shrugged and sighed. "O.K.,'' he said. "O.K.. So I am still optimistic. I have sharp eyes,'' he pointed toward his own right eye, which stared balefully at Busse. "I have sharp eyes. But I am optimistic. We will make steel by next May.'' He sighed. "When I was in Poland, I saw the Pope bless a rolling mill with holy water. You are going to need the Pope.''

■ ■ ■

Kris McGee, the manager of construction, operated out of an office in one end of a construction trailer among the Run-a-Muckers' trailers. In McGee's trailer I counted a total of forty-two different building permits hung

■

on the walls, displayed in little plastic frames. These were local and state building permits. Conspicuous by its absence was a federal building permit from the Environmental Protection Agency—a permit for the construction of a steel mill in the middle of Indiana. That building permit had not yet been issued by the EPA's regional office in Chicago, because the whole idea of putting a steel mill beside a small town in Indiana seemed controversial to the federal agency.

There were two potential sources of pollution from the Crawfordsville steel mill. The first was air pollution. There would be a slight but noticeable decrease in the quality of the ambient air downwind from the mill. Most of the air pollution would be gases and tiny smoke particles coming from the electric arc furnaces in the melt shop. Nucor would clean the furnace smoke by passing it through filter bags in a baghouse, a pollution-control building near the melt shop. The filter bags would remove zinc and lead particles before the smoke was discharged into the atmosphere. By the time the smoke was released it would be nearly invisible, but it would contain carbon dioxide, carbon monoxide, and oxides of nitrogen.

The second potential source of pollution was the steel mill's water systems. The water came from the wells dug under Cornstalk Creek. Most of it would be sprayed on red-hot steel, evaporating into the air at a rate of two million gallons a day. The water vapor was nothing but clean steam, because it was boiled from clean steel. But there were also closed water loops in the mill.

The water in the closed loops accumulated oils and chemical contaminants. The oils would be skimmed off and sold to waste-oil dealers. The chemicals would be filtered out of the water through lime and charcoal filters, and the lime and charcoal sludge would be hauled away and buried in landfills.

By far the largest source of pollution was indirect. The steel mill would consume more than a hundred million watts of electricity around the clock, to melt steel and drive the engines of the CSP. The electricity was generated by Public Service of Indiana in coal-fired generating stations that burned soft coal, a source of acid rain and carbon dioxide. Coal-fired power stations located in various places in Indiana would contribute energy to the mill, and coal smoke and gases would drift north by east into Ohio, Pennsylvania, New Jersey, and New York. Busse had got a sweet deal on the price of electricity from Public Service: he was going to pay 1.6 cents per kilowatt hour. The electricity bills would come to $1.5 million a month.

■

Nucor ran computer studies to try to predict the windborne dispersion of pollutants coming directly from the mill. EPA officials analyzed the Nucor tests and recommended changes in the design of Nucor's pollution systems. "Every time we tested, we found that our pollution-control technology was not good enough," said Busse. "We would change the design and resubmit it to the EPA. Then the government rules would change. That went on and on. It took forever, but it was done right."

While the haggling with the EPA was going on, Ken Iverson chose to let the Crawfordsville Project roll ahead full-speed without an environmental permit. Iverson and Busse believed that the Nucor Corporation was in a global race to perfect the thin-slab casting of steel, and could not afford to wait. In a technical if not a literal sense, the Crawfordsville Project was an unlawful act. Nucor ultimately would pay nearly a quarter of a million dollars in fines to the EPA for building the steel mill without a permit. The Nucor mill was reasonably clean but was crash-built, which offended the EPA.

■ ■ ■

"Half the time I step in a gopher hole, but once in a while a light bulb goes off in my head," said Keith Busse, as we drove in a pickup truck on a tour of inspection through the Crawfordsville Project. Speaking of gopher holes, he stopped the truck near a vast trench, two stories deep, that ran between the melt shop and the hot mill building, from south to north along the line of the CSP. The trench marked the roots of the CSP's tunnel furnace, 540 feet long. The tunnel furnace would link the casting machine (in the melt shop) with the CSP's rolling stands (in the hot mill building). The tunnel furnace was not built by SMS. It was a joint effort by an American company, Bricmont & Associates, and a French company, Stein Heurtey. The tunnel furnace was a sort of long oven that would soak the slab with heat before the slab was driven through the millstands to be rolled flat.

Busse pointed toward a concrete wall, two stories high, that sat at the bottom of the tunnel furnace trench. "The contractor's going to have to tear out that wall," he said. "He wasn't up to snuff on his concrete. He got caught. He sucked the hot egg."

A cellular telephone in the truck beeped. Busse picked it up. "This is Keith. Guy fell? Uh-huh. How far did he fall? There's an ambulance on its way?" Busse swung the truck around and gunned it.

A construction worker was lying at the bottom of a pit near the hot mill.

■

The man was conscious and in pain, with an injured back. A crowd had gathered.

Busse telephoned the Crawfordsville police department. "This is Keith Busse, at Nucor Steel. We've had an accident. We've got an ambulance coming through. We want a road cleared in one and a half minutes." Moments later an ambulance from Culver Union Hospital bumped across the job site, lights flashing. Busse parked his truck and jumped out. He found his safety director, one Janice E. Roach, who had already arrived at the scene. She told Busse that the construction worker had fallen twelve feet, that the extent of his injuries was not known, but that his back did not seem to be broken. Jan Roach was a pretty woman in her late thirties, with curly light-brown hair and sharp brown eyes behind glasses. Her hair flowed rather spectacularly from beneath a Nucor hard hat with a red cross on it.

The injured man was brought out of the pit on a stretcher and loaded into the ambulance, and the ambulance bumped its way back across the job site, lights flashing.

With that incident wrapped tight, Busse climbed back into his truck and drove it inside the hot mill, to check on progress with the lower half of the CSP. He drove the truck through a wide doorway into a huge space, and the truck lurched across a dirt floor. The hot mill building was 880 feet long, the length of three football fields. Far overhead, the roof of the hot mill sparkled with sodium lights in late afternoon. The air inside the building was hazy with dust thrown up by cherry-picker cranes that were working inside the building, and the building stank of diesel fumes and echoed with shouts of construction workers. Busse stopped the truck beside the four millstands of the CSP, the CSP's rolling mill, which would roll the steel flat.

The millstands sat in the center of a pit sliced through yellow Indiana mud. The millstands consisted of four pairs of roll housings, doughnuts standing upright, 26 feet tall. Each millstand would contain a set of rolls like rolling pins. The millstands rested on a kind of barge made of pre-stressed concrete. The barge sat in the center of the pit, and the pit was mostly filled with concrete bunkers, stairways, and tunnels.

"There's your average four-story building down in there," said Busse, indicating the pit. "If this plant was built by U.S. Steel, it would cost six hundred million dollars, easy," he said, leaning on the steering wheel of the truck. "We may end up doing it for less than two hundred and sixty million. And if this facility was built by the government, I don't know

■

what it would cost. A billion dollars, two billion, I don't know. And if I had a quarter for every time I was pissed off at the Germans, we could have built this mill for free.''

A rolling mill is like an iceberg, in that most of its mass is invisible. Steel rolling machines float on barges of concrete. Most of a steel mill's mass is below ground and out of sight. The tops of the machines, which may rise several stories above the ground, are the tip of the iceberg. The millstands of the CSP weighed a total of 1,000 tons in an unfinished state. They would weigh 2,000 tons, in total, by the time all their moving parts had been installed. The moving parts included rolls, bearings, and hydraulic machinery. The concrete barge, upon which the millstands floated, had been designed to support 2,000 tons of vibrating machinery without sinking into the soil. Nucor steel reinforcement bars stuck out in all directions from the concrete, like a pincushion. The four millstands were bolted tightly into the concrete with bolts the size of truck axles. They were bolted down because machines for shaping and rolling steel are subjected to tremendous shaking when red-hot metal rumbles through them, and yet they must be aligned to microscopic tolerances, or the steel will go out of whack and flutter as it moves through the millstands, and will come out with swollen surfaces and tattered edges. The four millstands had been lined up with each other, straight and true, to a tolerance of $\frac{1}{10,000}$ of an inch, equal to the thickness of $\frac{1}{50}$ of a human hair.

Apart from the concrete raft supporting the millstands, most of the floor of the hot mill consisted of native Indiana dirt, piled in heaps and cut with trenches, or covered with a web of reinforcement bars, waiting for concrete to be poured on it. The hot mill building was supposed to have a concrete floor by now. But the floor wasn't finished. Even though the CSP's millstands had been installed, excavation equipment was still digging holes inside the building. The normal way to build a factory requires things to be done in a certain sequence: first the shell of the building is erected, next concrete floors are poured, and then the machines are delivered and bolted to the floor. Keith Busse was not building his steel mill in sequence. He had taken delivery of the machines before the holes in the ground had even been dug. In fact, parts of the hot mill building had still not yet been designed. Busse was building everything all at once: fast-track construction.

One wall of the hot mill building was still open to wind and rain. Busse wondered out loud, ''What would happen if it rained torrentially for a

■

month and soaked out the ground around the building?'' He didn't know what would happen, and it worried him. The CSP might sink into the mud, might warp off a true line. The CSP, 1,000 feet from the casting tower to the downcoiler, was trued along a straight center line to a tolerance of not more than $^{20}/_{1,000}$ of an inch over the entire length of the machine—the thickness of four human hairs laid side by side.

■ ■ ■

The sun was going down and Busse wanted to get home for dinner. He drove the pickup truck out of the hot mill, parked it near the Admin Building, and went into his office. A window in his office looked out on a red sky broken by spidery outlines of erection cranes. Busse sat down at his desk and opened a calendar to the month of October, in which all thirty-one days were a blackened snarl of obligations. He jotted a few more obligations on top of obligations. "My wife, Carol, just doesn't understand," he said. " 'I never see you,' she says. 'But I *am* home every night,' I tell her."

He hardly had time to plink watermelons. "I called the gun store the other day and my partner said to me, 'We thought you had died.' I said, 'I'm not dead, I'm building a steel mill.' "

What worried him most of all was the undeniable fact that he did not know how to make steel. "Me, a hot metal man?" he said. "I'm just a man trying to learn. And right now, I know enough about steel to get out there and be dangerous as hell."

Busse sighed and reached under his desk, and pulled off his steel-toed boots. He kicked them aside and shoved his feet into a pair of burgundy tasseled loafers. While it made him nervous to think about the startup, he regarded his lack of experience with molten steel as an operational strength for the Nucor Corporation, because he was free to make his own mistakes and free to learn from them.

"I understand that at the Lake Shore, the engineers and operational people are terribly frustrated," he said. "They don't get a role in decisions. They are merely *given* a project. The board of directors and a few high-ranking managers of the company make all the decisions, and then those decisions are thrust on the operational people. The operators can't touch a mill while it's being built, but they are expected to make it *go*. And if the mill is a piece of junk, and you are the operator, you get your ass kicked to the moon! People at Big Steel are held so accountable they are

■

afraid of their own shadows. But management is never held accountable and is always on a witch hunt. It tends to take the integrated steel companies a minimum of five years to make up their minds to do anything at all. It's just comical to listen to these guys from Big Steel. The speed at which Nucor moves mystifies them, astounds them, and they're jealous of it. They're in a cover-your-ass mode. It's CYA, CYA, right up the line. Because a bad decision tends to cost you your job. People at Big Steel aren't rewarded for taking chances, even if they make a mistake.''

Busse put on a blue hopsack jacket and jerked his shoulders to settle it. ''I've got to visit the little boys' room,'' he said. He went into a rest room and stood before a urinal, washed his hands, crumpled a paper towel, and glanced into a mirror. His face—square cheekbones, fluffy hair, broad jaw, green penetrating eyes—looked back at him with a neutral expression. He left the Admin Building and went into the parking lot.

It was a cold, windy evening. A burnt October plain surrounded the Project. A green Nucor flag and an American flag snapped on flagpoles by the Admin Building. An old oak tree stood in a field near the parking lot, blown lopsided by the wind, and beyond the tree, a horned moon was climbing.

Busse climbed into his car, the white Audi sedan with smoked windows that concealed the occupant. He disagreed with those who might describe his car as white. ''It's not white, it's *pearl*,'' he insisted. Naturally he kept a rod in the glovebox—''I'm prepared to meet danger in any form.'' He started the engine and said, ''Just follow me.''

I climbed into a rented Nissan Sentra and started my engine. Busse swung fast through a parking lot and past the melt shop, which stood up from the dust in a half-finished state, missing some of its panels, but it was beginning to look like a melt shop—a tall structure framed with steel and covered with steel, a cathedral of steel. He drove past two ladles that were sitting outdoors and collecting rainwater, waiting for hot metal next spring, and he drove past a guard trailer. Then he turned east on a county road and pulled away from me at ninety miles an hour.

I floored my Nissan and the chase began. Two yellow clouds floated in repose over the half-finished steel mill, behind us, in a blue-black afternoon. We moved so fast that the earth seemed to display its fundamental curve, and the three buildings of the Crawfordsville Project fell away behind us.

At the hamlet of New Ross, Busse's car veered left and I nearly lost him. The speed at which he moved was mystifying and astounding. We

■

headed onto another county road. Downshifting through the village of Max, I watched for police. Beyond Max, I tortured the Nissan and caught a glimpse of a pearlescent gleam on the road ahead.

On one occasion (he once admitted to me) he had chosen to run from a police car. He wouldn't have done it except that the cops had chased him. Busse was a law-and-order man but he did not intend to lose *any* chase once it began, so he dodged under a bridge while the cops roared over the bridge and into the distance with their bubble-gum lights flashing.

As we passed through the next town, Busse slowed to let me approach, and then he took off again. It was a cat-and-mouse game. Each town had one grain elevator. Storefronts were often covered with plywood. The sun burned through bald timber on the horizon. Crows waddled through corn stubble, eyeing fallen grain. We entered the outskirts of Indianapolis, and suddenly Busse's car dodged through a brick gateway and we were in a development of manor homes.

Busse's house was five thousand square feet of newness done up with a hint of the pioneer style. The house featured stained cedar siding, a brick facade, and thermopane glass. Carol Busse, his wife, was preparing dinner in the kitchen. Spotlights in the kitchen gleamed upon floors of polished oak. Carol was a slender woman in her thirties, younger than Keith, with wavy blonde hair and a master's degree in education, who said that she did not like to cook. "I hope you like Chinese food," she said in a dubious voice. She had brought home a box of frozen shrimp rolls and she was frying them in a skillet. Keith and Carol's child, a nine-year-old boy named Aaron, romped around a den beside the kitchen. Carol was Keith's second wife. In addition to Aaron, Keith had four other children: two sons and two daughters by his first wife, Valerie Moellering. They were in their twenties.

Busse handed me a beer, popped one for himself, and we stood on a deck overlooking an empty back yard. Large houses were outlined in the dusk. The houses loomed over empty back yards planted with spindly trees supported by cords. Busse had planted trees on his property because, in his opinion, trees increase the value of a piece of real estate.

He sipped his beer. "I'm the only one from in-dus-try in my neighborhood," he remarked, pronouncing the word "industry" with a Hoosier firmness: *in*-dus-try, accenting the first syllable and elongating the word like a piece of flat-rolled steel. The word seemed to hang over the neighborhood: *in*-dus-try. "For some reason there seem to be mostly anesthe-

■

siologists and heart surgeons in this neighborhood,'' he said. "I think I'm the only one from in-dus-try.''

Carol Busse served the shrimp rolls on china, and we sat down at a glass table in the dining room. The dining room was decorated with silk flowers, in mauves and lavenders, growing from china vases. Aaron joined us at the table. We made small talk over the shrimp rolls. Aaron had played soccer today. Carol had had a rather tiring day with the soccer game and running Aaron around. Keith had had a rather tiring day, building a steel mill. The shrimp rolls disappeared, except for Aaron's shrimp roll, which sat on a plate under Aaron's trenchant eyes.

"I want some ice cream,'' said Aaron.

"You're going to get sherbet,'' said Carol, "after you finish your shrimp roll.''

"Mom makes us eat sherbet,'' said Keith.

Aaron finished his shrimp roll and we ate orange sherbet.

Keith took me downstairs to the basement of the house. He entered a strongroom and flipped on the lights. The room was Busse's chamber of the imagination. In the center of the room there was a pool table, and on the wall he had hung Winchester rifles. They were reproductions, fresh-minted to commemorate great events in the winning of the West. Each gun had a name, he explained, pulling off the wall a Winchester known as "The Little Bighorn.'' It had an engraved breech. "I've got the Antlered Game, the Golden Spike, the Buffalo Bill, and the Teddy Roosevelt. I don't have the Duke,'' he said in a voice of regret. "I can't afford the Duke,'' he said. "We are running out of frontiers. We have still not found a way to get to other planets, much less to other universes. We have done all the pioneering that ever will be done, except to build a city under the sea or travel to the stars.'' Trying to rebuild the American steel industry was not quite the same thing, to him; it was a flawed and diminished substitute for pioneering into a wilderness. He turned off the lights and I followed him to a wet bar in the basement, where he uncorked a bottle of red wine, poured it into glasses, and sat down on a couch to tell his personal story concerning the wars of steel.

■ ■ ■

Busse's father, Earl Busse, was a fireman in Fort Wayne. He worked at Engine House No. 4 in Fort Wayne for most of his life. "My father gave us a decent house, but he had nothing in his pockets to show for it.'' Earl Busse was promoted to platoon captain toward the end of his career.

■

The Busses went to Lutheran church every Sunday and gathered at the grandparents' house on holy days. At Christmas, the grandfather, August, unwrapped a silver music box that his father brought over from Germany in the late nineteenth century. It was a Christmas-tree stand. August fitted a fir tree into the music box, and Louise, the grandmother, lit real candles in the tree. (Earl, the fireman, objected.) The tree turned around and around, and the candles smoked and fluttered, and the silver music box played a rich music from the Old World that reminded Keith of something perfect and unattainable. The family sang "O Tannenbaum" in German. "Our idea of a big-time program was a sing-along," he said.

When Busse was a senior in high school, in 1961, he told his father that he wanted to go to the University of Indiana at Bloomington. "My father told me that he couldn't even afford to send me to a regional state college." No college for Keith. "I remember vividly that I cried when he told me that. I went across town to my godfather, Charlie Deak, and he loaned me seven hundred and fifty dollars for college. Charlie said to me that there were other ways to get a job done."

Busse enrolled at the International Junior College, in Fort Wayne, while he supported himself by pumping gas at a Montgomery Ward automobile service center. If a truck full of batteries arrived at the service center, he unloaded batteries until one o'clock in the morning. The battery acid ate holes in his clothes and his skin. He did his homework at the kitchen table at night, where he often fell asleep at the table, facedown on an accounting textbook. At five o'clock in the morning, his father would come into the kitchen to cook breakfast, and Keith would wake up to find the book glued to his cheek with drool. He peeled the book off his face while his father cooked him an egg.

With a junior college degree, he went into accounting as a career, and after working at accounting jobs with the McGill Manufacturing Company, Square D, and Dresser Industries, he ended up managing a janitorial supply business in Fort Wayne. The business, Moellering Supply, was owned by Busse's father-in-law, the father of Busse's first wife, Valerie Moellering. "He had no sons and he wanted somebody to leave his business to. The trouble was that Valerie and I did a lot of fighting at the supper table over her father's business. I yearned to get back into in-dus-try." Meanwhile Busse had been going to night school at Saint Francis College, in Fort Wayne, where he finally got a bachelor's degree in 1972, at the age of twenty-nine.

That same year, a headhunter called him to tell him about a steel joist factory that was being built twenty miles northeast of Fort Wayne, in Saint Joe, Indiana, by the Nucor Corporation. Busse took a job as division controller at Nucor's yet-unbuilt Vulcraft joist plant in Saint Joe. That was his first factory startup. "I got my tail on board Nucor and it's been a rocket ride ever since." The ride took him away from Valerie Moellering. "She never did forgive me for quitting her father's business. I was argumentative and so was she. She wanted to go to family reunions, and I was up at Nucor to the wee hours, working on budgets." Their marriage broke up three years after he joined Nucor. Iverson promoted Busse to be the general manager of the joist plant. Meanwhile Busse attended Indiana University at night school—finally reaching his goal to attend Indiana University. There he earned an M.B.A.

A couple of years after Busse became the general manager of the Vulcraft plant in Saint Joe, the industrial depression of the early 1980s turned the Midwest into the Rust Belt. "The joist plant at Saint Joe took it in the guts," said Busse. "We were operating at one-third capacity. We weren't all that necessary to Nucor at that time, and if we had folded, I don't know that anybody would have noticed."

Busse did not want to go down in history as the first Nucor manager to close a plant and fire the workers. He flew to Charlotte frequently, where he shouted at Ken Iverson, who got to talking loud with Keith Busse. "We fought for our people's interests in the marketplace and with Charlotte," said Busse. "We took a stick to other Nucor divisions that showed mediocre performance—we crucified mediocrity at Nucor. We kept the gates open at Saint Joe as long as we could, so that the employees could at least earn their base pay."

Busse's workers were working a three-day week, and their pay collapsed (it was based on production bonuses, and there wasn't much production), but nobody at Saint Joe was laid off. "Everybody put supper on their plate," said Busse. "Nobody had to pick up and go to Texas. And we never went into a loss during a month. There aren't many manufacturing plants in the *world* that can operate at one-third capacity, with all the employees kept on the payroll, and still make a profit." During the darkest month, the Saint Joe plant sold only 2,000 tons of steel joists, with a profit of a single dollar per ton of steel, for a monthly net profit of $2,000 for the entire factory. "You can't imagine what a thing of pride that two thousand dollars was for us," said Busse.

Busse put the factory's cash flow, such as it was, into research and development, in order to give his people something to do. Most manufac-

■

turing companies in recessions lay people off and cut back on research and development; Busse laid off nobody and increased his R & D. "We challenged our people to come up with new ideas."

His people, with nothing else to do, invented and patented a prefabricated steel roof. "It's the best steel roof that money can buy," claimed Busse. "We carried that monkey on our backs all through the recession."

Busse had hunched himself up on the sofa, and now he leaned forward with his elbows on his knees, and his green eyes and his blocky face grew hard and intense as he remembered the wars of steel, the toughest business on earth.

"The name of the game in steel is survival," he said. "We made ourselves into a crack unit. When we came out of the recession, we were razor sharp. We did combat, and we rolled over everybody." Ticking his fingers, he listed his enemies and their fates: "Socar hung on for a while and died. Ceco died. Republic Steel left the joist business. Macomber got out of the joist business and later died completely. Gooder Hendrichsen is a small firm in Chicago, and hanging on. Can-Am—that's a company up in northern New England—is doing well, by imitating Nucor. Berger Joist went into Chapter 11. When one of those eastern bastards like Standard Building Systems came west into our territory, we made them eat bar steel sideways. We killed most of our competitors, to tell you the truth."

Busse collected the customers of bankrupt joist companies like an inheritance. The Saint Joe factory complex—which by then included a steel decking plant as well as the joist plant—was being carried on Nucor's books at a value of $14 million. Busse's factory went on to earn a gross pretax profit of around $14 million in a single year—it earned in one year a gross profit equal to the cost of building the factory. That was a 100 percent pretax return on assets. Not bad for steel.

Keith Busse had a victory on his résumé. He had seen how much money a factory can make. You wouldn't believe how much money a factory could make if the workers were motivated and the manufacturing process was tuned to a hot blue flame of high-speed production. It was something the Japanese understood, and now Keith Busse understood it. He had seen the miracle of Industry.

Just before Christmas in 1984, the telephone rang in Busse's office in Saint Joe, Indiana. It was Iverson and Aycock, on a conference call.

IVERSON: Keith, we're having a project that we'd like you to undertake.
BUSSE: Yes, sir, what is that?

■

IVERSON: We're investigating Nucor's entry into the fastener business.
[*Silence on the Busse end of the line.*]

AYCOCK: Bolts, Keith.

BUSSE: It's awfully close to Christmas. You guys've been drinking and you stuck your fingers on the wrong digits on the telephone! This is Keith Busse at Saint Joe, Indiana. I don't know jack about bolts.

The chairman informed Busse that Nucor had been approached by Stanadyne, an American manufacturing company, to form a joint venture to manufacture bolts. Stanadyne owed a mothballed bolt factory in Chicago, and Stanadyne proposed uprooting the machines and moving them to a Nucor steel mill, where nonunion labor would make the bolts. "It sounds like a good deal," said Iverson to Busse. Would Busse like to come to Chicago with Iverson to take a look at these bolt-making machines?

Iverson and Busse toured the Chicago bolt factory. Busse didn't like what he saw. The machines were antiques and they operated slowly; they could make less than fifty bolts per man-minute of labor. He told Iverson, "Forget about pulling ancient machines up by the roots and moving them somewhere else. You've got to build from the ground up, Ken. We're going to compete with Chinese labor at fifty cents an hour. We can only do it through technology."

So Iverson dispatched Busse to the Pacific Rim to gather the latest intelligence on bolts. Busse went to Tokyo and Osaka, to Nippon Steel and to Kobe Steel, keeping his eyes open, and he went to Taiwan, where the Taiwanese bolt manufacturers were impressed to hear that Busse was in the firearms business. He traded stories about guns for tips on bolts. He saw bolt-making machines in Asia that could outstream the bolts at a rate of 1,600 bolts per man-minute. Talk about a high rate of fire, that was 26 bolts a second, faster than a Magic Dragon on full auto. Upon his return from the Pacific Rim, he went to the Cotswold Building, where the conversation went like this, according to Busse:

"So what do you think?" said Iverson.

"The question, Ken, is what shape the bolt factory will take," said Busse. "The American bolt makers got into a bad habit of sending the bolt out for heat-treating and chrome plating, after they had made the bolt. Then they trucked the bolt *back* to the factory to be packaged. So if you make a bolt the American way, you are trucking your bolt all over God's green earth. You've got to put all three bolt-manufacturing processes into one building, or you are gonna lose your shirt."

■

Iverson's reply went something like this: Look, I called you with a cheap proposal to buy a few old machines. And you said it wouldn't work, and so I sent you to Japan. "And now you're telling me we've got to start three new businesses!"

Iverson's secretary hurried to close Iverson's door.

"That's absolutely right," said Busse. "Three businesses: making your bolt, chrome-plating it, and packaging it, all in one building. That's what I'm telling you, Ken. Or you're gonna have your shorts ripped off."

"How much money are we talking about?"

"Twenty million. And bolts are a skinny business."

"What's the annual return?"

"Only eighteen percent," replied Busse. "Not at all the kind of return on capital that this company has come to expect in steel."

An eighteen percent annual yield on an investment didn't sound too bad to Iverson. That was a lot better than a money market account. Iverson said something like: O.K., go talk to Sam Siegel, he'll give you twenty million for bolts.

And then Bethlehem Steel closed the last standard-bolt factory in the United States. The Bethlehem bolt plant was a cluster of scorched brick buildings in Lebanon, Pennsylvania, that had once been the largest bolt factory in the world, employing 2,400 people. For decades Bethlehem Steel hadn't invested more than chump change in the plant, and now Bethlehem planned to close the factory and fire a couple thousand people. Bolts were of no further interest to Bethlehem Steel, they were only a commodity where the profit margins stank. All of a sudden, Nucor was going to be the only major bolt maker in the United States.

"It was an unnatural situation for all standard bolts to be imported," as Busse put it. "It's Ken Iverson's view that when you have an unnatural situation, you also have an opportunity to make a lot of money." With Bethlehem Steel out of the bolt business, Nucor was staring at a chance to cut a large slice of the American bolt market.

Back in the Rust Belt with a capital budget for bolts, Busse sat down one Saturday afternoon with a Nucor manager by the auspicious name of Wayne Studebaker. On that Saturday afternoon, Busse and Studebaker designed the first major American bolt factory to be built in perhaps thirty years. They drew the factory's floor plan on sheets of plain white paper, without help from outside design consultants. They could not have hired any design consultants anyway. There was nobody left in America who knew how to design a bolt factory. Keith Busse was the only American expert on bolt plants.

■

He and Wayne Studebaker drew the factory largely from Busse's memory of what Busse had seen in Japan. It was a technology transfer, done from memory, not unlike the technology transfers that abounded in Japan during the 1960s, when Japanese managers toured American factories with their eyes open, carrying little notebooks in their hands. The Japanese managers had gone home and studied their notebooks and sketched from memory what they had seen, adding their own ideas, trying to leapfrog the Americans. Busse thought that it would be nice to put robots on his factory floor to carry bolts around the building. Let robots truck your bolts all over God's green earth. And then Busse started digging a hole in God's green earth in Saint Joe, Indiana, next door to the joist plant, for the Nucor Fastener Division, the bolt factory. Eleven months later, the factory was finished and farmers were running the robots.

Today, one out of eight standard bolts in the United States comes from Keith Busse's bolt plant. But the bolt plant isn't making any money for Nucor. The bolt plant is running at near-full capacity and is barely breaking even. Busse's factory is outstreaming the bolts at full auto into the guts of the American market and earning no money on bolts.

"We were a bunch of ignorant bastards," said Busse.

Much to Nucor's surprise, the *Canadians* turned out to be the tough competitors in the bolt business, not the Japanese. "It's those little fellows up in Canada that are giving us fits," said Busse. "Those little bolt guys in Canada don't want to give up market share. So they lower the price of bolts, and we lower the price of bolts, and they lower the price of bolts. We're locked in a vicious gun battle for market share. It's going to be a struggle and a heartache, but I think we will prevail in bolts."

"What will you do if you don't prevail?" I asked.

"If a thing doesn't work, we tear it out and start over. That's what we do at this company."

Busse wasn't finished with factories. He persuaded Iverson to give him $12 million to build a factory to make prefabricated metal buildings. Busse built this factory in the middle of cornfields in Waterloo, Indiana, ten miles from Saint Joe—the Nucor Building Products factory. The factory makes prefabricated steel buildings for small factories—in other words, it's a factory for making factories. Iverson figured that as the Rust Belt revived, there would be a continuing demand for factories. The Nucor factory buildings feature Nucor's patented steel roof, the best steel roof that money can buy, which Busse's people invented during the recession with the R & D money that Busse had taken out of the joist plant's withered cash flow.

■

Nucor Building Products turned out to be the same story as the bolt plant, all over again. Nucor isn't making any money on metal factory buildings. "We're locked in another deadly marketing struggle," said Busse. "And at the beginning, we were not up to snuff on quality. Bold and arrogant Nucor. I would characterize our attitude as that of the South in 1862. We rode gallantly off to war and came home in tatters. The market clubbed the daylights out of us."

So far, both of Keith Busse's factories were candidates for a tear-out, although the Nucor Corporation and Keith Busse remained officially hopeful. Keith Busse had displayed a certain flair for the crash project, for the industrial wing job, but his new factories had not brought home profits, and profits, after all, are what factories are all about. At virtually any company other than Nucor, an accountant who had gone over his head in bolts, robots, farmers, and a factory for manufacturing factory buildings would have been finished, dead meat. Companies have methods for dealing with any Keith Busses that happen to crop up unexpectedly in the organization: the guy is given a bum's rush to the door. A press release chases the guy out the door, announcing that he has left the company "to pursue opportunities elsewhere."

Busse's factories had created a delicate situation within the Nucor Corporation. David Aycock characterized the problem this way: "A failure is a particularly dangerous time for a company, and must be handled with great skill. If people at higher levels come down on an employee and say, 'You goofed up,' it can kill all initiative at a company."

"So how do you handle someone who does goof?" I asked.

Aycock, the ex-welder not unfamiliar with mules, put his hands flat on his desk and eyed me in a way that suggested he thought I was an idiot. "You give 'em something new to do!" he said in a flaring voice. "Because they're the only damn people in the company who dared to take any risks! I'm talkin' *personal* risks!"

So they told Keith Busse to go ahead and build the first big American steel mill in more than twenty years.

■ ■ ■

Busse had stopped going to Lutheran church. He and Carol had gone to a local Methodist church until Busse had become the chairman of the church's finance committee. "That was a mistake," he admitted, because it had led to what he described as "a gigantic falling-out" with the Methodist elders over their handling of money. He felt that the Methodist church

■

was spending its money "on copying machines and office furniture," as he put it, just like a corporate bureaucracy. And so he quit the church out of an indignant mixture of moral, management, and accounting principles. He had lost, or had never been able to find, the simple family life that he had known as a boy, when the old people gathered at holidays to play cards at Grandpa August's house in Fort Wayne, and listened to Grandpa's silver music box at Christmas, and sang songs in German. When Platoon Captain Earl Busse died, a hearse carried the casket past the largest engine house in Fort Wayne, Engine House No. 1. The firemen brought their red machines into the street and stood at attention beside them as Earl Busse went by, a rare honor for an ordinary fireman who rose, slowly, to the level of platoon captain. Earl Busse died with little in his pockets, but he had friends in Fort Wayne.

"What I miss so deeply today," said Keith Earl Busse, "are the sing-alongs at Grandpa and Grandma's. For some reason I have never been able to get it together with my family and have a sing-along. The idea of a sing-along seems to have vanished. I don't believe we'll ever have another sing-along." He leaned back on the sofa in his basement and slapped his knees gently. "I have to get up at five tomorrow," he said.

■ ■ ■

Fall turned into winter, winter turned into spring, and Nucor sank a quarter of a billion dollars into the mud of central Indiana. The money went down with a great turmoil and vanished, like a submarine blowing its tanks for a polar voyage. The Project continued to creep over budget. Iverson had originally planned a construction budget of $225 million; that had jumped to $250 million, and then slipped up to $260 million. The cost overruns began to scare Busse—not that they were so large, but he didn't know where they might end, and they were heading toward a third of a billion dollars. Even $260 million was a lot of money. If $260 million were a mass of $10 bills, and the $10 bills were laid end to end on the ground in a straight line, the money would stretch from Charlotte, North Carolina, to Seattle, Washington.

As he watched the long green stream march away before his eyes, Busse felt the inexpressible grief of a sharp-pencil man who cannot stop negative cash flow. He wanted to start up this steel mill in the worst way. He told Iverson that he could have the ancient Austrian machines in the cold mill rolling steel by January of 1989. Since the Compact Strip Production machine wasn't anywhere near ready to make steel, Busse planned to

■

purchase coils of semifinished hot-rolled steel from one of the big boys up on the Lake Shore, cold-roll it in Nucor's cold mill to put a shine on the metal, and then sell it for a slight profit, just to get some cash flow coming into the plant.

But the German blueprints for the used cold-rolling machines became a nightmare for the Run-a-Muckers. Nobody could read the prints, because they were written in German. What was even worse, about half of the plans had been lost in Germany during the past twenty years. It was like putting together Tinkertoys, consisting of hundreds of thousands of parts, with the help of partial instructions written in German. Busse eventually hired a German translator, who studied the prints and supplied English words for things like *Obere Andrückroll, H 19 Spreizhasp Spannzylinder, Anti-Crimproll* . . . and gradually the cold mill machinery was bolted together and wired up. But after that, nothing much happened inside the cold mill, except that a lot of software crashed.

That winter, a Red River man by the name of John O'Hair was standing on a roof joist, on the framework of the tunnel furnace building, holding a piece of the finest steel roof that money can buy, a Nucor roof panel, when a gust of wind came along, flipped the panel, and pushed him off the joist. He wasn't clipped in. John O'Hair screamed on the way down. His internal organs burst on impact and he died.

Winter snowed itself out and the winds of April 1989 breathed on Crawfordsville. Along with spring came tornadoes. A funnel cloud passed over the Project. It gave off a deep, continuous roar, like a herd of bison. The funnel licked down like a whip, and stung the roof of the cold mill, throwing pieces of Nucor steel to the winds—a reminder that steel is vanity, steel is grass. Nucor blamed its problems on the Westinghouse Electric Corporation.

Westinghouse was the main electrical contractor for the Crawfordsville Project, the contractor in charge of the "electrics," as computers, software, and motor-drive systems in a steel mill are called. Westinghouse was having trouble with its software in the cold mill. The Westinghouse man who had been writing the software that would run the antique machines suffered a heart attack and died abruptly. It was a disaster. The new Westinghouse software expert couldn't figure out his predecessor's half-written programs.

Nucor's irritation at Westinghouse mounted when Westinghouse ran into problems with the electrics in the millstands of the CSP. Some 2 million feet of electrical cables (380 miles of cables) inside the CSP went into a hairball. A subterranean rage from the bowels of the Nucor Corporation boiled up at Westinghouse, until some Nucor steelworkers began to refer to the Westinghouse engineers as "the circle jerks"—referring to the Westinghouse corporate logo, a circle surrounding a *W*. But Westinghouse was one of the few corporations in the world that had the expertise to untangle Nucor's wires when Nucor could not untangle the wires by itself. Nucor's electrical problems, and Nucor's dependence on Westinghouse to solve them, drove Keith Busse crazy.

"The Westinghouse people are putting in a three- or four-day workweek," he complained. "They arrive from Pittsburgh in the middle of a Monday and leave early on a Friday. There isn't a single one of our people who can fully understand the electrics at this mill. *Westinghouse* can't understand 'em!"

Westinghouse, headquartered in Pittsburgh, was eleven times the size of the Nucor Corporation. "Dealing with Westinghouse is like reading the riot act to the Empire State Building," said Busse.

By spring, Mark Millett and his people and various contractors, unhampered by electrical problems, had installed the SMS thin-slab casting machine within the concrete buttresses of the casting tower, and had installed and wired up the electric arc furnaces. One day Busse declared to me on the telephone that Millett was getting ready to melt steel and pour it into the casting machine. "I swallow an alligator every week," he said. "If I swallow the last alligator, and Mark Millett swallows his alligators, we'll strike an arc in June."

"Have you set a date for the first cast?" I asked Busse.

"You obviously have not been through the startup of a steel mill," he replied. "Neither have I. So I can't answer your question. I have no idea when we'll cast. I know *what* will happen. Once Mark has done a test melt in a furnace, the people on the casting deck will get so anxious to cast a heat they won't be able to sleep. They'll start working around the clock. After that, the first cast could come at any time. Whenever they feel they're ready, they'll pour a heat in the machine. The first cast could come at three o'clock in the morning, when it comes. Starting up a casting machine is like having a baby. The steel comes when it's ready to be born."

PART II

STARTUP

11

A CRISIS

■ ■ ■

I flew to Indiana in June, on the brink of the startup, with a change of planes at the municipal airport in Dayton, Ohio. The Dayton municipal airport is a coronary artery in the Rust Belt. There I noticed a steel salesman making calls to prospects at a pay telephone. This guy was a life-giving corpuscle. He wore a panther-link gold bracelet and a blended suit the color of the moon-washed night sky. He put a quarter in the phone. He covered one ear with his hand. The bracelet jinked on his wrist. "How are ya?" he said into the receiver. "Good. The ingots I've got for you are sort of bastard stuff. There's chrome in the ingots, but not enough chrome. And do you still want the pig iron?"

Prospect would take the bastard ingots but couldn't decide about the pig iron. "No problem, I'll talk to you later." He moved along.

If a steel salesperson works on commission he or she earns around four dollars a ton for your plain carbon steels. That's a commission of two-tenths of a cent per pound. If you think about trying to make a living by selling anything on a commission of two-tenths of a cent per pound, you begin to understand the life of someone in steel sales. The name of the game is survival. The name of the game is tons.

I boarded a flight to Indianapolis. From Indianapolis I drove to Craw-fordsville and down Washington Street, past the Montgomery County Courthouse, past a line of Victorian houses, past a line of bungalows, past a cemetery, past the town's McDonald's, and into open country on the Old Ladoga Road, where the landscape opened into a feeling of possibilities.

■

The Old Ladoga Road traveled south by west through occasional timber and long stretches of farmland, dipping and rising among cornfields. In early June, the corn was as high as a dachshund's eye. Hedges of wild white multiflora roses were creaming between the fields, filling the air with a delicate fragrance. The road passed near little farmhouses buried in wind-slanted trees. The wind blew trees into teardrops, and the skies were changeable.

I turned east on a county road and crossed the headwaters of Offield Creek, a brushy gurgle that drains into Sugar Creek. The road climbed a hill and cut across the old Monon Line, and the road traveled along a barbed-wire fence beside a line of railroad cars heaped with macerated automobiles. The road turned past a guard house, where a sign read, "Warning: Patrolled by Armed Guards," and led the road into the Crawfordsville Project.

■ ■ ■

The three buildings of the Crawfordsville Project were finished—three featureless oblongs as big as blimp hangars, covered with Nucor steel panels and painted a tan color: minimalist rectangles, but in their bulk they hinted at the presence of monstrous industrial engines within them. The melt shop and the hot mill, the hangars of the CSP, stood at right angles to each other, in an L joined by the tunnel furnace building, the middle stretch of the CSP. The cold mill brooded alone, a rectangular structure eight hundred feet long by five hundred feet wide, large enough to cover four or five football fields. You could see a hole in its roof where the tornado had lashed it.

The three buildings floated in a sea of grass that gave off a tingly smell. The grass was wheat, green and young, mowed as tight as a fairway. It looked like a lawn but it would die at the end of summer.

On concrete sidewalks that floated in the shaved wheat, I saw men and women hurrying here and there. For the most part they were young and wore green Nucor hard hats and either blue jeans or blue workclothes. Each of them had passed the Nucor psychological examination for self-reliance. The trailers of the Run-a-Muckers sat crumpled, beaten, and locked. The Run-a-Muckers had been disbanded and sent to their stations in far-flung parts of the plant. The construction trailers were mostly closed, and Kris McGee, the manager of construction, was nowhere to be seen. A pair of boots stuck out from the back of a construction trailer, a contractor taking a nap.

■

At the Admin Building, Keith Busse leaned back in his chair and looked at me with green restless eyes. "I'm not on an even keel right now," he said. "I want this startup to be *done with,* and it's not even begun."

He had gained weight; he had stabilized his uneven keel with ballast. The telephone rang. "Yes? So we've now concluded that the motor on the caster is inadequate?"

I noticed a hand grenade sitting on a shelf near his desk. A sign on the hand grenade read, "Complaint Department. Please Take a Number." A tag with the number *1* was tied to the hand grenade's pin. If you pulled number one, the pin came out of the grenade. There were no other numbers available.

Busse hung up the telephone and informed me that I had walked into a crisis.

"We've encountered terrible problems with the oscillating motor on the casting machine," he said. "SMS is telling us it'll take eight weeks to order a new motor. Meanwhile we can't operate the casting machine. So because of one little dinky motor, *we can't make steel.* Do you know how much an eight-week delay could cost this corporation? Eight million dollars. We're losing one million dollars a week."

These were operating losses, not capital costs, and they had to be reported as cash losses in the annual statement. So one blown motor might lose the company $8 million in 1989; or 40 cents a share in losses. One inadequate little motor could drag down Nucor's stock on the New York Stock Exchange. The Crawfordsville startup was undoubtedly going to take a terrible toll on Nucor's earnings, and this motor just made it worse.

"Every single day for the last five months it's been like this—we've missed another schedule, we've screwed up another deadline, we've blown the cost side of things! Putting piers in the wrong place. How could we be so stupid as to put piers in the wrong place! You write it off. And I have to go home to Carol every day in a positive frame of mind. It's tough to receive bad news all day and then be a bubbly guy with your family. How did we ever think we could do this mill for $250 million? We've spent $264,017,000 *as of this moment.* And my current capital budget is $265 million. So I have a capital budget with less than a million dollars left in it! You can see what's going to happen—the heartache of another cost-bust. We'll do this mill for $268 million, *maybe.* But when you start out with a budget, you say, 'Hey, I can beat this sucker!' " He slammed a fist into his palm. "Don't get me wrong. There are a lot of people out there in

■

the steel industry with their jaws hanging open. They're saying to us, 'How did you ever do this mill for under half a billion?' "

"What are they saying about you, Keith?"

"They're probably saying I don't know shit from shine-ola about making steel."

He sighed and placed one ankle across the opposite knee and threw his arms behind his head. His office smelled faintly of Calvin Klein's Obsession for Men, his cologne of choice. On his left wrist he wore a stainless steel Rolex Oyster Perpetual Date-Just watch, with a black face. He turned his gaze out the window, toward a row of spindly wind-tortured oaks. He had recently planted the saplings in the shaved wheat, since trees make a piece of property more valuable. The sky went through beautiful transformations in Busse's window. Clouds were rolling by in cloudstreets. The clouds knocked against each other, their gray crops bulging with rain.

"It's going to come hard," he said, and sighed. "Ken Iverson knows that. Ken has lived with each Nucor startup, some nineteen of them now. Ken knows. Dave Aycock knows. We tend to bullshit ourselves at Nucor. Like with bolts. We would never have got into some of these businesses if we had known the suffering we would encounter relative to profits. This mill could be the same story as bolts. This mill could be another struggle and a heartache. But I can't say that to Mark Millett. It would be a terrible mistake to destroy Mark's optimism. But for Mark and my other managers to confront so many unknowns out there . . ." Busse's voice trailed off, and he stared out the window at the clouds and the young trees. "We stuck an awesome responsibility on a kid," he said. "At the present time, Mark Millett is the most disturbed kid on the planet."

■ ■ ■

The most disturbed kid on the planet had recently moved into a small office attached to the melt shop, where I found him talking on the telephone.

"Super!" he said. "Excellent!" He hung up the telephone. "Hell," he said, sounding abruptly weary. "We've got a problem with the motor on the caster. It's the crisis of the day." A loop or two of disheveled brown hair hung over Millett's forehead. His hair had begun to show threads of gray. The carpet in Millett's office was plastered with yellow mud.

He picked up a green Nucor hard hat, passed a hand through his hair, and plopped the hard hat down on his head. "The formation of the gray hair is accelerating," he remarked. "Come on, I'll show you the caster."

We headed from east to west down the length of the melt shop, passing first through a repair shop. Blocks of machinery stood on the floor and against the walls. These were spare segments of the SMS thin-slab casting machine. SMS had built several identical casting machines with interchangeable modules. The modules would be drenched with liquid steel regularly, because there is no such thing as a startup without breakouts. After each breakout, sections of the machine would have to be lifted out of the casting tower, chipped with cold chisels, and repaired.

We picked a route through the spare casting machines to the casting tower, which stood near the east end of the melt shop.

The casting tower, the top end of the Compact Strip Production machine, was the center of the Crawfordsville Project. It stood at the place where the melt shop and the hot mill came together in an L, at the corner of the L. Liquid steel would crystallize inside the casting tower. The tower was located at the critical point of the manufacturing process. The casting tower resembled the bridge of a ship. It was a broad structure, standing thirty-two feet off the floor of the melt shop. Millett ran up a flight of steel stairs to the middle of the casting tower. There a pair of steel doors stood open to reveal the guts of the German casting machine, exposed under floodlights. The lights illuminated the vertical train of paired rolls, the rungs of the ladder through which the steel would travel downward in a ribbon.

You could look in two directions from the casting tower, along both shafts of the L, either northward along the axis of the Compact Strip Production machine, or westward up the axis of the melt shop. The view along the CSP extended along the tunnel furnace, through the narrow tunnel furnace building, and into the hot mill, where you could barely discern the CSP's millstands, the parts of the machine that would roll the steel flat. No portion of the CSP was visible beyond the millstands. It is not possible from any location at the Crawfordsville Project to view the Compact Strip Production machine in its entirety. And Iverson's machine is alleged to be a *miniature* steelmaking device.

"You could say I'm in the middle of the white water," Millett remarked, contemplating the view along the CSP.

He turned and ran up a flight of stairs to the casting deck, taking the steps two at a time. The casting deck formed the top surface of the casting tower. It was a concrete platform, much like the casting deck at Buschhütten, Germany, on top of Kolakowski's pilot machine. The casting machine extended below our feet, below the casting deck. I followed Millett to a slit in the center of the deck: the mouth of the copper funnel, the Kolakowski mold. This was the eye of the needle, because the mill's entire production

RICHARD PRESTON

of steel would have to pass through this hole. The mold was embedded in
a steel plate that stood twelve inches off the surface of the casting deck.
The plate covered the top of the casting machine, and it was painted green.
A double array of steel piers, ten feet tall, surrounded the mold opening,
the mouth of the machine. Wide sets of tracks ran on top of the piers. The
tracks guided ladles over the mouth of the machine.

Millett peered down into the Kolakowski mold, looking twenty feet
down inside the machine. "It looks endless down in there, doesn't it?" he
remarked.

I ran my hand over the inner faces of the copper mold. The copper was
a reddish gold in color, cool to the touch, and polished as smooth as glass.

Millett led me to the west railing of the casting deck, and we looked
westward down the axis of the melt shop. The melt shop was a cavernous,
murky space, 130 feet high—the interior height of Reims Cathedral. The
building was 850 feet long—longer than a cathedral.

The melt shop contained three stations, or decks, tall structures on
concrete piers, crammed with complexes of machinery and electrical gear.
The decks were the casting deck, where we stood; the metallurgy deck,
located near the center of the melt shop; and the furnace deck, which
extended along the west end of the melt shop. The building extended
westward beyond the furnace deck into a scrap bay, where railroad cars
would deliver the Rust Belt to the melt shop. The two electric arc
furnaces—the north furnace and the south furnace—were embedded in the
furnace deck. The floor of the melt shop was bare Indiana earth. Mobile
structures known as overhead cranes were moving up and down the length
of the melt shop, along crane rails at the roof of the building. The crane
rails and the roof were supported by the long double row of Duane Hurler's
columns, marching into the distance. Tiny human figures crossed the floor
or gathered on the decks in knots. Brilliant pinpoints of arc welders glittered through haze and smoke, as welding crews made last-minute changes
to girders and pipes—the final tear-outs. The view opened into the sort of
dreamlike, inexplicable profundities one glimpses in Giambattista Pirane-
si's "Carceri," the etchings of infinite torture chambers.

"After two years of work, it's good to see all your screw-ups come to
life," remarked Millett, leaning on the railing of the casting deck, crossing
one steel-toed boot over the other. He was twenty-nine years old, and in
charge of one of the most complex works of technology anywhere in the
global steel industry. "It's a pretty little melt shop, if I must say so
myself," he added.

■

"Why do you call this building a 'shop'?" I asked him.

" 'Melt shop'—that's a term favored by the minimills," he answered. "People from U.S. Steel would call it a 'steelworks.' I guess that word's a little too grand for Nucor."

"Keith Busse called this place a pimple on the camel's butt," I said.

"Which camel is Keith referring to?"

"Big Steel," I said.

He grinned. "If you don't treat a pimple it can become a boil."

■ ■ ■

Mark Millett ran down a flight of stairs, taking them two at a time, sliding his hands along the rails for balance, his boots thudding on the stairs. We entered a small dungeon buried inside the casting tower. The dungeon was strewn with tools and it stank of basso profundo sweat. A clump of Nucor steelworkers, local guys, churned and shouted around an electric motor that sat on a pedestal in the center of the room. It was the dead motor, the crisis. The motor was supposed to make the Kowlakowski mold jog up and down, to prevent steel from sticking to the mold.

The steelworkers were whacking the motor's spindles with a hammer and a piece of wood. They wore blue workclothes and green hard hats and most of them wore gold chains around their necks—single-herringbone chains, double-herringbone chains, and bright flat oval-link gold chains, which flashed in sodium light. Flies zipped through the dungeon and swarmed over the men.

"So wot's happening, mates?" said Millett.

"A beer'd be real good right now," said a steelworker.

An engineer by the name of Patrick Owens came over to Millett to discuss the situation. Owens was the commissioning engineer for the casting machine. He worked for SMS Concast, Inc., an American subsidiary of SMS. "We're going to give this motor one more test, Mark," said Owens. "It'll surprise the shit out of me if it works." He dragged on a cigarette and looked over his shoulder. "Again!" he said to the steelworkers, and they hit the motor with the hammer and wood. "Again! Again!"

Big Dave Thompson, casting foreman, clambered into the dungeon. "Where's the pry bar?" he rumbled. "The motor's gotta go south a couple of inches." Thompson bent over the motor, tinkering with it.

Two SMS engineers appeared, Germans in white hard hats. The Germans put their white hard hats together and had a confabulation in the

■

German language. They announced to Millett that there was nothing wrong with their motor.

The dungeon was viciously hot. Somewhere down in the guts of the machine a pipe grinder wailed, filling the dungeon with a smell of sparking metal. A cutting torch popped with a blue flash.

"Let's go," said Owens, the commissioning engineer. He unclipped the microphone of a walkie-talkie from his shirt and shouted, "Start it up!"

The motor hummed and shook, but the drive shaft did not rotate.

"We never moved!" shouted Owens to his walkie-talkie.

The steelworkers offered comments:

"It's because we don't have enough horsies in that motor."

"Well, go to the Mustang Ranch and get a few. Those horsies'll get your motor going."

"Nucor's got the technology, but the reason none of it works is the people are a bunch of lunatics."

"Yeah, lunatics off the farm," said Big Dave Thompson. He jerked his thumb at a steelworker. "We got you right off the farm here in Crawfordsville."

"Hey! Don't lay that farmin' on me! I've never been farmin' a day in my life!"

Millett said, "We're going to put in a bigger motor, whether we need it or not. That's the way we're going to go." Two white German hard hats nodded, slightly. But still, it would take eight weeks to get a new motor.

Nucor maintenance men thought they could do better than that, and they made some telephone calls, and eventually they located a bigger motor of the right type. It happened to be attached to the old Hazelett casting machine in Darlington, and the motor was immediately shipped by truck to Indiana. But the problem wasn't solved yet, because an oscillator motor on a casting machine is controlled by a twenty-thousand-dollar computer the size of a microwave oven known as a motor drive. The Nucor people couldn't find the right sort of motor-drive computer to drive the new motor. They called all around the United States, and discovered that the vendors of such computers—General Electric and Siemens—were out of stock. It seemed that all of the known computers of this type were already hooked up to casting machines operated by Big Steel companies. This was a bad piece of news. It was obvious that no Big Steel company was going to sell Nucor one of those computers, not for any price. Plans were made to steal a computer from Big Steel. The only solution was theft. Otherwise the casting machine wouldn't be able to cast steel for weeks, if not months.

■ ■ ■

It is evening. The location is a booth in the bar of the General Lew Wallace motel, in Crawfordsville. A Nucor maintenance man, making phone calls, has discovered that a flagship American steel mill on the shore of Lake Michigan owns a computer of the right type sitting unused on a shelf.

This Nucor maintenance man is a heavyset fellow with horn-rimmed spectacles. He is smoking a Winston and nursing a Budweiser out of a long-necked brown beauty. "I'm a maintenance guy, and I've got a friend up on the Lake Shore who's a maintenance guy," he says. "I called my friend. When maintenance guys get into trouble they take care of each other. My friend on the Lake Shore is a fraternity brother. I can probably get this motor drive from him without the blessing of his company, but he's risking severe discipline for stealing if he gets caught. So first he's going to ask permission from his boss."

"And if his boss says 'no'?" wonders Millett.

"Then I wouldn't be above taking that thing surreptitiously." The maintenance man tips his beer to his mouth.

Millett says, "It's interesting, this subterranean cooperation between Nucor and Big Steel."

"Isn't it, though?" says the maintenance man.

■ ■ ■

With the Crawfordsville Project stopped dead in its tracks and bleeding money by the hour, some quiet contacts proceeded through the Lake Shore steel mill that owned the computer, while Keith Busse sweated. There followed a telephone conversation between a Big Steel operations man and one of Busse's people. The Big Steel man was a shirt-sleeve type who worked at a mill, as opposed to a vice-president from Corporate. He personally offered to help Nucor. Nucor wouldn't have to steal the computer from his mill. He could have the computer delivered to Crawfordsville within forty-eight hours, at no charge, absolutely free. Tell Keith Busse not to worry. He didn't need to add that Busse now owed him a favor.

The Big Steel man was a little nervous about being caught in this fifth-column operation, and cautioned the Nucor manager not to tell anyone where that computer came from. Corporate would put his tit in the wringer if they found out he had "loaned" a twenty-thousand-dollar black box of electronic gear to help Nucor start the Crawfordsville Project. He

■

complained that it had begun to dawn on Corporate that Iverson's machine might actually work, and corporate vice-presidents at his company had suddenly become all hysterical on the subject of Nucor, calling meetings to worry about Nucor, burying him with memos on the subject of Nucor. His words, as they were reported to me, went like this: "This company's management is real bitter toward Nucor, at the corporate level. At first Corporate was pooh-poohing Nucor's technology. They now regard Nucor as a real threat. Now every time I meet with Corporate I have to hear about *the Nucor threat.*"

Two days later, a truck from the Lake Shore pulled into the Crawfordsville Project. Nucor maintenance people quickly unloaded a brown unmarked cardboard box from the truck, and the truck turned around and left. The box showed up at the casting tower. Nucor's electricians cut open the box and pulled out a spanking new custom-built computer, a tangle of wires and chips in a steel shell, with an LED display. It was packed in Styrofoam. It came with no return address, no packing slip, no instructions, and no warranty. They bolted the computer to a wall and wired it up to the casting machine and tied it to another computer. They entered a few commands on a keyboard, and the Kolakowski mold began to heave up and down gently, with a sound of breathing. The casting machine was beginning to act like a casting machine, with the help of Big Steel.

12

THE SCRAP PILES

■ ■ ■

Mark Millett stood on the casting deck and stared dubiously down into the mouth of the Kolakowski mold, into the depths of the casting machine, at water gushing over the machinery, at shadows of men moving across machinery. "Hm," he remarked.

"Yo! We're gittin' more water!" shouted a voice from a grotto under the casting deck. A wrench clanked on a pipe.

"I don't know where this God damned water is coming from!" shouted another voice, while steelworkers clattered and banged around inside the casting tower.

Millett decended two flights of stairs to the base of the casting tower, where he put his hands in his hip pockets and stared up at a waterfall pouring down through the casting tower. The water smelled rank and moist, like a cave.

"This is Nucor's contribution to the art world," he said.

I asked, "If you have water flooding over steel inside the casting machine, won't that cause an explosion?"

"No," he replied. "That won't do it. There's a saying in the steel industry that you can put water on steel, but you can't put steel on water. As long as the water is on top of the steel, it boils off. When there's water *underneath* steel, it turns to steam and blows things into the sky. It can go through the roof of a mill."

"Are you serious?"

■

"Yup. If you poured steel onto wet ground, it would go through the roof of the mill."

He explained that a dirt floor is preferable to a concrete floor in a melt shop, because dirt can be kept dry as a bone, while concrete wicks up moisture like a sponge. Any moisture trapped in concrete can flash to steam under a steel splash, and the concrete explodes beneath the steel, throwing concrete shrapnel mixed with liquid steel all over the place.

You can put water on steel, but you can't put steel on water. A human being is 70 percent water. A lot of the rest is carbon. Although a human being is mostly water, a body drenched with steel does not explode. Steel clings to human skin and burns downward into the flesh at Fahrenheit three thousand, seeking the carbon, letting off a barbecue smell of burning fat, hair, and meat.

At Nucor the steelworkers have certain polite terms for being hit by a splash. One is said to be "initiated" or "christened." They say to each other, "Have you been initiated?" The recipient of a heavy christening may run in circles, screaming, on fire. The steel soaks into one's clothes and they burn off in a flash, and then it eats into the skin. The victim may sometimes try to pull the steel off his skin, burning his fingertips off. Steel dissolves carbon and so it dissolves people.

■ ■ ■

The financial analysts on Wall Street were telling their clients that the success of the Crawfordsville Project depended on Nucor's ability to control the quality of the scrap steel that went into the furnaces. If high-quality junk went into the mill, high-quality steel would come out of it.

The man in charge of Nucor's scrap heaps was a foreman by the name of Jim Hoskins—an Alabaman in his late twenties, who until he had migrated to Crawfordsville had been a welder in a Vulcraft joist plant in Alabama, on an hourly wage. Hoskins had light-brown hair, a mustache, and a broad, kindly face, and he was a religious person.

Jim Hoskins's scrap piles were strung out for a mile along Nucor's railroad tracks, to the north and west of the melt shop. He took me for a walk one day down the length of his scrap piles. The piles were twenty feet tall and they stretched along the railroad tracks until they turned a bend and went out of sight, heading for Nucor's junction with the Monon Line.

"There's only about twenty-thousand tons of steel here," said Hoskins, with his foot on a heap of crushed rust and his eyes squinting down the tracks. "That's nothing. That's a one-week supply, if we're running the

■

furnaces at full steam.'' He walked a hundred yards and stopped by a pile of cubes made of shiny crushed steel. They resembled sugar cubes. They sparkled in a dull way under a cloudy sky. The steel cubes weighed one or two tons apiece, depending on how tightly they were packed, internally. The cubes, said Hoskins, were known as No. 1 bundles.

"No. 1 bundles are cubed-up stampings from a factory,'' he said. "These ones came from a General Motors truck and bus plant. So we call 'em truck and bus bundles. We've also got what we call Chrysler bundles and we've got Ford bundles. We're also going to be buying a test load of Isuzu bundles. I think we'll like the Isuzu steel fine. Later we'll get bundles of tin cans, and we'll see how they work in the furnaces.''

Hoskins kicked his boot along the outskirts of a house-sized pile of flat shards of steel cut into lacy shapes. "These are No. 1 bushelings out of Danville, Illinois,'' he said. "Bushelings are clippings and punchings from a factory—the insides of toasters, electric-motor pieces, stuff like that, good clean sheet steel.''

He stopped at an immense heap of rusty little fragments of steel. "This is shredded cars,'' he said. "See the transmission parts, motor parts, leaf springs, seat belt buckles, tie rods, gears, door handles?''

The pile glittered with door handles. They were chromed half-ovals. These were old cars. A piece of steel is nine to twenty years old by the time it re-enters a furnace; the recycling loop moves like a glacier.

About seventy-five percent of all steel in the United States is eventually recycled. The scrap dealers in the Midwest accumulate automobiles in piles so large that the piles are really iron mines. Mountain ranges of cars have been thrust up in the Rust Belt. Scrap is one of America's natural resources. America is the world's largest exporter of scrap. Americans load it into freighters and ship it overseas for a profit. The United States dominates the world markets for scrap. Scrap steel made in the U.S.A. is said to be the world's most desirable junk. American scrap is scrap of the first water. Other nations buy it, resell it, barter it, and pass it along. It goes to Turkey, Korea, Japan, Venezuela, and Italy. About one-third of all scrap steel traded among nations is American scrap. The United States exports ten million tons of scrap steel a year. We may not be able to export a memory chip to save our lives, but we can always flog off our ruin to Japan, where they melt it and turn it into Toyotas.

The scrap dealers have a machine that eats cars. It has twin hammer knives that can shred a car in thirty seconds. Scrap workers remove the car's motor, its lead battery, and some of its copper wiring before they

■

send the hulk into the hammer knives. They also rifle the hulk with their hands, feeling the glove compartment, the floormats, and the cracks between the seats, looking for loose change. They find an average of $1.50 to $2 in loose change per hulk. A big scrap dealer can shred 600 hulks a day. That yields $900 to $1,200 in coinage per day, or $300,000 to $400,000 in loose change a year—free cash flow.

The hulks fall out of the hammer knives as fist-sized pieces of steel, mixed with seat stuffing, glass, knobs, rubber chunks, and wires, all mixed up and moving together along a conveyor belt. The fragments must be sorted. The conveyor belt dips through a water tank, and the plastic and other light material float away, while the metal and glass stay on the conveyor belt, still mixed with steel. Next the conveyor belt emerges from the water and passes by human hand-pickers wearing gloves, who pick away chunks of chrome, copper, glass, aluminum, and brake shoe nuggets, until the conveyor belt carries almost nothing but chunks of plain carbon steel.

Scrap workers rifling their hands through hulks no doubt turn up the occasional twenty-dollar bill, pint of bourbon, set of dentures, diamond ring, gun. Anything that travels in cars travels in hulks. Maybe once in a while they find a mummy in a rotten gabardine suit, curled up in the car's trunk, with a hat worn low over its crispy face. If Jimmy Hoffa was put into a hulk he's due up any year now—the recycling loop proceeds like a glacier, and someone who falls into a crevasse in a wrecking yard may not emerge for fifteen years. If scrap workers do find a mummy I wouldn't be surprised if they were to handle it with the strictest professionalism—rifle the pockets for loose change, check the fingers for rings, and send the stiff into the hammer knives. In a case like that, the bones and shoes of the victim would dribble from the hammer knives, to be floated away in the water tank or picked away by the hand-pickers.

Factories and office buildings are dissolved in the hydrologic cycle of junk. When a skyscraper is torn down, an alligator shear, mounted on a hydraulic arm, clips pieces of structural steel out of the building and deposits it in trucks, and the building is trucked away. Same goes for bridges. Heavy machinery is sheared into pieces with a hydraulic shear, loaded into railroad cars, and taken to a steel mill. Freighters and supertankers are sliced into plates by demolition artists with torches, and the plates are carried away in railroad cars. When the railroad cars wear out, they are cut up with torches and piled into railroad cars and taken to meet their maker. Railroad wheels are a premium specialty in the scrap trade, a

■

gourmet item to be popped into a furnace like shallots, to add a *je ne sais quoi* to the soup.

Hoskins, the Nucor scrap man, was learning how to judge a pile of scrap steel the size of a house by looking at the color and texture of the metal. "If you're around the stuff long enough, you'll pick up on it," he said, somewhat hopefully. We stopped at the base of a mountain of filthy scabrous pieces of steel. "This is No. 2 scrap," he said. "It's a lower grade. There's everything and its brother in there—wheel rims, barbed-wire punchings, whole busted motor shafts, structural I-beams from buildings. Look—there's a roller from a steel mill." The roller looked like a rolling pin, except that it was the size of a tree trunk. Nucor ate steel mills for lunch.

A storm was brewing. Flashes of lightning cast an electric arc light over the piles of scrap, and in a matter of minutes, a black cloud ballooned over the steel mill. A wind came from the bottom of the cloud and began to whistle through the scrap piles. Green mammary lumps appeared on the cloud's underbelly, the sign of a tornado, and pieces of the cloud tore off and began to turn around and around, threatening to form a rattail. "We better get out of here," said Hoskins. We ran for the melt shop and got inside its doors just as the storm hit and a wall of water came down.

■ ■ ■

At the south furnace deck, Mark Millett stood at a railing and looked out through a scrap bay door, watching the sky strike arcs. In the distance a lone grain elevator in Whitesville faded away in the rain. Millett had met with Keith Busse that day, and he had told Busse that the arc furnace was ready to go. They had planned to tap the first heat of liquid steel tomorrow.

"If all goes well, we'll wake up Whitesville at sunrise," said Millett.

At Millett's feet, below the furnace deck, a magnetic crane picked up a clutch of No. 1 bushelings from a railroad car. The scrap dangled in glittering chains under the magnet. The crane dropped the scrap into a clamshell bucket, and the scrap crashed into the bucket with a sound like breaking glass. The clamshell bucket would be used to charge the furnace with scrap.

"Have you ever been inside an arc furnace before? No? Really? Well, we ought to do something about that," said Millett. His tone of voice meant, How perfectly unacceptable that you have never been inside an arc furnace. "Come on, then," he said. He led me across the furnace deck to the south electric arc furnace.

■

The furnace was a one-hundred-and-fifty-ton-capacity arc furnace designed by Asea Brown Boveri, Ltd., a Swiss multinational company that makes electrical equipment. The furnace looked like a cook pot, and it was embedded in the deck. It had a roof that functioned like a lid on a cook pot, to trap the heat when something is cooking inside. The lid was known technically as the furnace's roof. The lid could be moved sideways, to open the pot, so that scrap steel could be dropped into it. The furnace's roof, or lid, had three holes in it, for the insertion of the electrodes to melt the scrap.

Three graphite electrodes, twenty-five feet long and two feet in diameter, hung vertically in a triad from three electrode arms that extended over the furnace. The electrodes were as large as the boles of mature ponderosa pines. The electrode arms, which gripped the electrodes, were attached to a turret. The turret sat beside the furnace. The electrode arms could be lowered or raised. The arms could insert the electrodes down through the three holes in the roof of the furnace and into a huge pile of scrap metal sitting inside the furnace.

A concrete transformer vault sat next to the furnace, containing a single transformer, and from the vault came a hank of gross, fat, shiny electrical cables, hanging loosely and running to the electrode arms. The cables could deliver ninety million watts of power to the electrodes—about one-eighth the power output of a commercial nuclear reactor. Each power cable was a foot in diameter. They were water-cooled power cables. Water lines ran inside the cables to keep them cool. Without such an internal cooling system, the power cables would melt. When both arc furnaces were running at full power, they would together suck the equivalent of a quarter of a nuke out of the Indiana power grid.

There was a door in the furnace called the slag door. Millett jumped into the furnace through the slag door. His voice boomed out of the furnace: "Watch your step."

There was a gap in the furnace deck between the deck and the furnace. I inspected the gap. A dizzying view opened downward to the floor of the steel mill. It was a long way to the ground. I jumped across the space and landed on the lip of the slag door, and climbed into the furnace.

Millett stood in the center of the furnace, his feet resting on a layer of black granules. "We're standing in coke," he said. "I put a layer of coke in here. That's to alloy the steel with carbon." The furnace vessel was twenty-two feet in diameter, the size of an average living room, and circular. It was lined with a crumbly gray material, like stucco. Millett picked a piece of stucco from the wall and rolled it between his fingers. "This is gunning material, a type of refractory," he explained. "We sprayed it in here to insulate the walls, for the startup."

He explained that the furnace was cooled by water that circulated in pipes embedded in the furnace's wall panels. The wall panels consisted of a solid network of serpentine water pipes. "Well—" He looked around the furnace. "This is the quiet before the storm."

Millett wore a red, white, and blue polo shirt with a heraldic crest on it showing the lion and unicorn of the British Empire. His hands were filthy and his face was smudged with soot. He pulled a small piece of graph paper out of his hip pocket and jotted a reminder on it with a stubby pencil. The paper was wrinkled, smudged, and covered with illegible reminders. Millett did not wear a wristwatch, and he often asked people what time it was. "I don't like the idea of watching time," he explained. He had lost every wristwatch he ever owned. They seemed to fall off his wrist. He wore a silver chain around his neck, and on it hung a tarnished silver Saint Christopher medal, the patron saint of travelers and possibly of hot metal men, who seem to go anywhere for a heat of steel. "I don't *think* I could live in England," he said. "There are many aspects of England that I love, and some basic aspects that I don't like. But I retain my British passport. The one thing I can't do in America is vote."

"Mark Millett! Mark Millett!" a voice crackled inside the furnace.

Millett pulled a walkie-talkie from the hip pocket of his jeans. "Yes?"

Troubles in a water main.

Millett leaped out through the slag door, hurried across the furnace deck, descended a flight of stairs, and ran up another flight of stairs, tugging at the rails with both hands. Quickly he arrived at the metallurgy deck, in the center of the melt shop. There, an anxious group of engineers and Nucor foremen peered into a water trough, which was gurgling and foaming. Pipes were banging like a radiator.

"Ahoy!" said Millett. "Wot's happening?"

"There's air trapped in here, Mark. The water's coming back on us."

"We'd better shut it off," said Millett. The water tank overflowed. "Interesting," said Millett. He turned to a foreman. "Stick your head in that tank."

"O.K.," He grinned. "Why?"

"Why not?" said Millett.

The tank erupted with a groan, throwing water over Millett and the foreman. They jumped backward, laughing. Drenched and dripping, Mil-

■

lett ran down a flight of stairs to the floor of the melt shop, taking the steps two at a time, his hands sliding down the railings. A knot of anxious foremen and engineers hurried after their boss. Millett stopped at a water main and put his ear to it, listening to the water flowing through the main.

"We think the problem is in there, Mark," said a foreman. "There could be a hard hat trapped inside this line. That's a distinct possibility."

Millett put his hand on the water main, feeling for vibrations, and remarked that finding hard hats inside water mains was not uncommon when you had unionized construction workers building a nonunion steel mill. "We've found two helmets in pipes already," he said. "Of course, when you think you have a hard hat in a water main, your other thought is that the hard hat could be attached to a body."

They had crash-built the steel mill, and maybe some poor guy had been welded inside a pipe, and now he was a bloater moving through the system. Millett raced back up the stairs to the water trough and began opening and closing valves in random order, listening for sound effects. A great rumble shook the foundations of the melt shop. The foremen and engineers glanced at Millett nervously.

"We may have proverbially screwed something up," said Millett in a worried tone of voice.

"What now?" said a foreman.

"Hm," said Millett. He took off his hard hat and ran his hand through his hair.

"You know, Mark, I bought Nucor at forty-six," remarked an engineer.

"It's now at fifty-four," said Millett.

"I know," said the engineer. "My wife, she wanted to buy Disney stock. I said to her, 'One Mickey Mouse outfit around this house is enough.' "

"If this place takes off, Nucor's stock is gonna take off," said Millett.

"The stock hasn't done much yet," said the engineer. "I think a lot of investors are hanging off on the stock, waiting to see if the casting machine works."

"Yeah," said Millett.

He hurried back to the arc furnace and fiddled with some valves there. "We could have an air lock in the pipes," he remarked. He finally decided that the water level in the melt shop's water system was too high. He

■

170

ordered the water level to be lowered, and that seemed to reduce the banging in the pipes.

Millett wanted to charge the furnace full of scrap, to get that job over with, so that first thing tomorrow morning he could plunge the electrodes into the furnace and begin to melt steel. The clamshell bucket had been filled with scrap. An overhead crane picked up the clamshell bucket, moved it over the open furnace vessel, and dumped seventy tons of broken steel into the furnace with a slither and a crash.

An iron fog the color of cinnamon burped over the lip of the furnace, cascaded down the walls of the furnace, slid across the furnace deck, knee-high, and then the cinnamon cloud collided with a wall and climbed it, then slowly diffused, until it filled the west end of the melt shop with airborne rust. A red heap of metal trash poked above the lip of the furnace, decorated with a generous sprinkling of General Motors truck and bus bundles, piled like diced fruit. A white powder like confectioner's sugar was mixed with the pile. The powder was limestone. It would form a slag when the scrap was melted.

Millett walked in circles through the cinnamon fog, around and around the arc furnace. "Ahoy! What time is it? Clear off the deck! Get everything off the deck that doesn't belong up here." Pulling a handful of purple wires out of a control panel, he remarked to his walkie-talkie, "I've pushed every button on this panel, and it still isn't working." Later to his walkie-talkie: "We're ready to go. We should run this furnace right now. By tomorrow we'll have screwed up something else."

■ ■ ■

Millett stood on a dock on a lake in Indiana, with a gin and tonic in hand, an Englishman in the Rust Belt, consoled with quinine. The first heat was scheduled to begin tomorrow at first light. Mark and Abby Millett's house was a brown-brick contemporary, with pointed gables, on a dirt road. Their house looked west over the lake, among ash trees and beeches.

"We'll strike an arc tomorrow at six A.M.," he said. "We'll try to get the furnace on fire right away. If all goes well, we'll have a finished heat of steel by eight A.M. But I wouldn't put a dime on that." He grinned, half to himself, and sipped his gin and tonic.

A breeze wobbled the beech leaves, exposing their pale undersides. Funnel clouds had been touching down all over the state that afternoon,

■

and a twister had just detonated a few mobile homes in DeLong, in north-eastern Indiana.

"I'm not pessimistic, but you've got to be prepared," he said. "I've tried to prepare, I've tried to think about which way the steel may go, if there's an accident."

The lake was gray-green and translucent. Millett peered into the water. "There's a lot of catfish and bass in there. One of these days I'll fish for them."

He climbed a flight of wooden steps up to a patio, where there was a grill heaped with charcoal briquets. He poured an ominous amount of Gulf lighter fluid over the briquets, circling the spray around and around until the briquets glistened, and struck a paper match. Flames leaped up. He rubbed singed hair off the back of his hand, his face glowing in a petroleum wildfire.

In the living room, Abby Millett was playing with their two small children. Abby was a slender person, in her late twenties, with curly dark-blonde hair, blue eyes, and a delicate face. She wore a dress and deck shoes. The children, Kate and Drew, ages two and one, respectively, had pale white-blond hair and the distracted faces and grubby fingers of toddlers. Mark sat down on a tan rug, and his daughter ran across the rug toward him, screaming with delight, and fell into his arms. The room was modestly decorated, with tan couches and a white-brick fireplace.

Mark cooked dinner while Abby put the children to bed. He removed a couple of pork loins from the refrigerator, carried them to the patio, and draped them over the grill. Back in the kitchen, he peeled potatoes.

"I barely have time to mow the grass these days," he remarked. "I get a half a day off a week. I said to Abby, 'When I'm home, all I do is cut grass.' Just maintaining the vegetation. That's about all I can handle right now. Children tend to happen. Small bodies can't relate to the fact that I'm not home when they have to go to sleep. Abby is a tough girl. Eight years ago in Colorado we had no responsibilities, nothing to do but ski."

He had finished peeling the potatoes. He cut them into chunks and spread the chunks in a pan. He dug a kitchen spoon into a brick of lard, and dropped blobs of lard all over the potatoes, *glap, plap,* and pushed the pan into the oven. "I hope you like English roast potatoes. We ate them in my family." He adjusted the heat in the oven. "The countdown starts at eleven P.M. tonight," he said. "That's when we start preheating two ladles. A ladle has to be hot before you can pour steel into it. If you had a cold ladle, the steel would cool off too much when you poured it into the

■

ladle. Also the brick lining of the ladle would crack. A couple of guys will come into the melt shop tonight and turn on gas burners, inside the ladles. They'll babysit the ladles 'til morning, 'til they're hot. We'll tap the first heat into a ladle at twenty-nine hundred degrees Fahrenheit.''

They made him a *soupcier* at the ski lodge, and his specialty was cream of broccoli soup. He cooked it in a ten-gallon kettle, but for some reason he couldn't control the flames properly and often scorched his soup. Frequently he had to dump ten gallons of soup down the drain, and every time that happened, he got into a dispute with the management over his cooking skills. Millett began to experiment with additives to cover up the stench of burned broccoli and to get it down the throat of the customer without turbulence, without causing the customer to vomit or jam up. He discovered that if he dumped a sack of white sugar into ten gallons of burned broccoli soup, the sugar obliterated the taste of burned broccoli. His treated soup resembled molten green cake icing mixed with black jimmies, but it met with customer acceptance. The customer had no idea he was eating a high-carbon soup alloyed with white sugar.

Millett was just mixing us each another gin and tonic, when suddenly he cried, ''The pork loins!'' He grabbed a wooden board and rushed outdoors to the patio. He returned in a moment with a grave look on his face and with two objects on the board, the pork loins. They were black blow-torched cylinders, gasping tendrils of smoke, as shiny as anthracite coal.

Millett placed the board without comment on a kitchen counter and opened the pork with a knife. ''Hm, yeah,'' he said, inspecting the muscle. He enlarged our drinks with extra gin.

Abby returned to the kitchen, having put the children to bed. She remarked, ''When we first moved into this house, we discovered that our neighbors already knew us.''

''Yeah, we were pretty weirded out by that,'' said Mark.

''It's a lot different than New York,'' said Abby. She chopped a green pepper and added it to a bowl of salad.

''It's another world, actually,'' said Mark. ''It's a good place for our kids to grow up in. Here they won't be out on the streets, beating on heads.''

He placed the burned loins on a serving plate. He put a knife to them and sliced them cross-grain into chips ringed with black. He peered into a little saucepan on the stove, shook it, and stirred it with a spoon. The saucepan contained a flecky green oobleck that looked suspiciously like it might feature burned broccoli. ''I don't have a recipe for this,'' he explained. ''I sort of made it up. I call it, 'A Sauce.' " He held the pan over the pork and laid a green drool along it, hurrying out A Sauce with his spoon.

A tallowy smell was coming from the oven. He opened the oven door and removed the pan of English roast potatoes. They had gathered into a sunken breakwater lashed by a foaming surf of lard. I looked sidelong at Abby Millett. No sign of concern in her face. At least Mark's timing was good. At least he had the parts of the dinner ready at the same time, and now all one had to do was eat them.

With potholders he carried the pan of potatoes into the dining room and placed it on a trivet on the table. He dimmed the chandelier. He brought the meat and salad out to the table, along with a boater of A Sauce, and we sat down to eat.

The pork loins had a sinister crust, but the crust was unaccountably sweet. It held the juices inside the meat. The pork was tender, succulent, lightly aged, offering savory overtones of garlic, sage, and salt. Millett's oobleck was a *sauce de fines herbes,* fragrant with fresh mint. He had picked the mint in his dooryard garden. The English roast potatoes were crisp on the outside, flaky on the inside, and happily they had not absorbed the lard. The salad was as honest as the month of June. I began to consider the possibility that Millett might succeed at cooking plain carbon steel tomorrow.

13

ARC LIGHT

∎ ▇ ∎

T he day of the first heat dawned rainy and cool. Mark Millett drove his
rust-colored 1979 Pontiac Grand Prix through the gates of the Crawfordsville Project at five o'clock in the morning and docked the car
with a rumble near the Run-a-Muckers' vacant trailers, in gray light. He
joined a knot of furnace men as they emerged from a locker near the melt
shop. They hurried into the melt shop, picking their way around puddles.

The men wore "furnace greens," spark-proof green suits, a common
sight in steel mills, and they wore hard hats with dark furnace glasses
attached to the brims. They passed under the piers of the casting tower and
into the long nave of the melt shop. A roar of gas jets filled the melt shop
with a low rumble. Near the casting tower there was a small deck, six feet
off the ground, where two ladles were propped on their sides, exhaling
flames. This was the ladle preheat station, where the ladles were being
heated red-hot to accept steel. Some of the furnace men hung around the
ladles, talking with the ladle preheat crew, and others hurried down the
melt shop, passed underneath the metallurgy deck, and climbed three
stories to the south furnace deck.

They stood around with their hands in their pockets, staring at a green
cook pot embedded in the deck, the Brown Boveri arc furnace filled with
seventy tons of No. 1 bushelings out of Danville, Illinois, topped with
cubed General Motors truck and bus scraps. They stared at the electrode
arms that held the three carbon electrodes, now positioned off to one side
of the furnace. They stared at their feet. The furnace crew consisted mostly

∎

175

of local men, although the foremen were experienced melters. The men kept putting on and taking off their gloves.

At an electrical substation beside the melt shop, an electrician named Marilyn Hendrickson threw a switch and a row of aluminum bars rose into place with a clack, and the system was energized. On the furnace deck, a chemist named Jeri Switzer laid out a row of cardboard sticks, thermocouple probes. She was a laboratory technician, and it was her job to keep the furnace men supplied with probes to test liquid steel.

The furnace men piled up a supply of eighteen-foot steel pipes. The pipes were oxygen lances, used to blow oxygen into liquid steel. Janice Roach, the safety director, handed out foam earplugs. A cold rainy wind whistled through an open bay and across the furnace deck. The steelworkers hung a wide blue plastic tarpaulin across the bay, to keep wind and rain off the deck.

Mark Millett stood by himself, his back and elbows resting on the rail of the furnace deck. He wore a red checkered lumberjack shirt. He described it as his most comfortable shirt, good for a startup. He believed he would do some sweating before the sun went down. A lone whistle pierced the air.

"What time is it?" said Millett.

"Five-thirty," replied a steelworker.

"Incredible how many people it takes to start the first heat," said Millett. "In a month, there won't be anybody up here except a few melters." He pulled his walkie-talkie from his back pocket and keyed up an engineer at the plant's water pumping station. He said, "O.K., you can start the water."

High-pressure water pumps surged, driving coolant water into the arc furnace's serpentine panels with a long wicked ramping hiss. Then it turned into a whine: *wheeeeeeeee!*

People looked around. Either an Exocet missile was heading for the melt shop or there was something wrong with the water.

Wheeeeee, boom! Crack! A thunder of roaring water echoed below the furnace deck.

Millett raced to the railing and stared downward. A water main had cracked, and water was pouring all over the dirt floor.

Millett hadn't cured all the water problems.

"Son of a bitch!" said a furnace man, taking off his gloves.

Millett's walkie-talkie had gone hysterical: "What happened? What was that?" it shouted.

"We broke a valve," Millett replied. "It looks like the water hit a closed valve. That was a pretty big bang. We're lucky we didn't blow anything else. A bit of a disconcerting start. We'll need to get a work crew over here to replace this valve. Ten-four."

A work crew moved a cherry-picker crane up to the water main. Steelworkers began to tinker with the valve, with wrenches.

"Well, we're not going to have a heat before eight A.M.," said Millett.

"What time you think we'll tap the heat?" someone asked him.

Millett shrugged.

Keith Busse was nowhere to be seen. Apparently Busse didn't want to crowd Millett today.

At 7:20 A.M. the water main was repaired, and Millett ordered the water pumps to be restarted. Again there was a hiss, but this time no bombs fell. As soon as Millett was satisfied that the water was circulating normally through the furnace, he gave an order to ready the electrodes for a meltdown.

At a control pulpit in a corner of the furnace deck, a Swiss engineer from Brown Boveri threw a switch to close the roof of the furnace. The roof moved sideways until it covered the pot. Then the three electrode arms moved over the roof of the furnace, until the three electrodes were poised above the three holes in the furnace's roof.

Millett fished into his pocket and screwed foam earplugs into his ears. He nodded to the Swiss engineers. "Directly!" he said.

The electrode arms moved downward. The electrodes slid into the furnace. There was a clunk. Then total silence. The arms withdrew the electrodes from the furnace.

"Damn! Damn! Damn!" a furnace worker muttered, taking off his gloves and pulling out his earplugs.

Millett went inside the control pulpit to confer with the Swiss engineers.

A rumor went around the deck that a breaker switch had accidentally tripped, shutting off power to the electrodes. The Swiss engineers opened cabinets inside the pulpit and ripped out a spaghetti-pile of wires, examining them.

Millett came out of the pulpit and signaled for another attempt at a long-arc meltdown.

The electrodes descended into the furnace. *Clunk.* Silence.

The Swiss engineers dragged wires all over the floor of the control pulpit, mixed them up, spoke in Swiss German to each other, toggled switches, and scratched their heads. At 8:45 A.M., the electrodes were again lowered into the furnace. There was a blue flash. A fierce concussion broke the air. The sound of the startup hit like a slap across the face.

■ ■ ■

A velvety brown mushroom cloud burped from the furnace and filled the west end of the melt shop with smoke. The furnace hummed and banged and gave off a loud buzz, and then it began to throw sparks over our heads in long soaring yellow lines. I ducked. I looked around. People's mouths were open. They were either screaming in fright or cheering or both. I couldn't hear them.

The electrodes burrowed into the scrap at 65 million watts of power. This was no halfhearted warm-up, Millett had goosed the throttle. The furnace was pulling a high tap on a long arc, as they say in the world of electric steel. At the flip of a switch, Millett had added a city to the Indiana power grid. The furnace was tapping enormous amounts of electrical energy out of the state's power grid, reaching to within spitting distance of its maximum draw, and driving the energy at high voltage into seventy tons of broken steel, enlacing the pile with arcs of electricity. These people didn't fool around when they started up a mill. The noise, a battering crashing rumble, could have cracked a molar. It made you want to get out of the building.

For five minutes the furnace pulled a high tap on a long arc. Abruptly the furnace shut down. The electrodes withdrew from the furnace, sliding back up three throats they had drilled in the pile, and when the electrodes appeared over the roof of the furnace, their white-hot tips glowed too brightly to look at with the naked eye. A group of furnace men lowered their dark glasses and approached the electrode tips and studied them, their expectant upturned faces bathed in light, their eyes black disks. The electrode tips dimmed to yellow and then to cherry-red.

Millett huddled in the pulpit with the Swiss engineers, studying the screen of a Compaq computer. They tore more wires out of a panel. Millett left the pulpit and walked in circles around the arc furnace, with his hands in the back pockets of his jeans, studying the serpentine water pipes in the furnace, looking for water leaks, a symptom of a pending steam explosion. Arc furnaces could explode. They threw out a burst of shrapnel followed by a tidal bore of molten steel.

The arms lowered the electrodes into the furnace. There was another earsplitting boom, and the electrodes shook and wobbled as they reopened three throats in the scrap. The furnace deck trembled steadily. Sparks

■

blossomed out of the furnace and sailed through the melt shop, bouncing across the deck and through the crowd. There were about twenty people on the furnace deck, and they watched without speaking, ducking sparks.

The banging and crashing inside the furnace smoothed out to a steady drumroll. Huge quantities of smoke and fire boiled from the furnace. Most of it was drawn off by an exhaust duct, but some smoke and fire gushed out around the furnace's lid. The furnace gave off a crumpling, crackling, unvarying string of booms that never faded away. A spectrum of low-frequency waves pulsed in the roar, vibrating the whole building, as if the melt shop were filled with transparent wiggling gelatin.

The electrodes pushed downward through the pile, enlarging the three throats in the heap of steel, burrowing downward into the steel at a rate of eight feet a minute. Arc bursts—sparks—extended from the tips of the electrodes straight downward into the steel. At sixty-five megawatts, a high tap on a long arc, the arcs were a foot and a half long and the diameter of a rope. They were blue plasma ropes, whipping around inside the throats drilled by the electrodes. The arcs, the blue ropes, lashed across steel, the steel boiled, caught fire, foamed, vaporized, and exploded. The arcs literally hurled scrap away from the electrodes. Heat from the boiling metal penetrated the pile of scrap, and the seventy tons of truck and bus bundles and toaster-metal punchings began to warm up and emit smoke, like a municipal dump on fire. Fluid steel collected in a puddle in the bottom of the furnace. The puddle rose and began to drown solid pieces of steel in a bath of hot metal.

The roar carried an under-buzz of a sixty-cycle AC hum, like a cheap loudspeaker. Concussions pressed on my body until they seemed to fill it, shaking the mastoid bone, chest, neck, forehead, and feet. The hum vibrated my larynx. I felt that my throat was humming by itself. It was a hum-along with electric steel. Occasionally the furnace lapsed into silence—the arcs were burning into a big piece of iron—and then the object blew apart like popcorn, a thunderclap shook the deck, and another mushroom cloud ascended to the roof. Low-frequency sound waves fluttered my diaphragm, and the loudest pops from the furnace produced a light, gentle, involuntary thoracic shudder. I sat on a bag of chromium sand, cringing at sparks.

An arc furnace can produce electrical disturbances in a state's power grid, echoes or harmonics that back-ripple through the grid. The furnace talks to the grid. Nucor mills were equipped with protective gear to dampen harmonics, but typically, during a startup, the gear fails to work

■

properly, and when Nucor pulls a high tap on a long arc in a new furnace, there is a drop in voltage throughout the entire state, and the power chatters and wavers as if someone is running a monstrous vacuum cleaner off the power grid. The chattering is strongest in towns near the steel mill and near the ends of utility lines, in outlying parts of the grid. It can cause fluorescent lights to blink and television sets to flicker and computers to crash.

The furnace workers wondered if folks in central Indiana would miss their soap operas this morning. They wondered if Keith Busse would get a telephone call from Public Service of Indiana, telling Busse to shut down his furnace, every town between Wallace and Raccoon is up in arms. The telephone call never came. If anyone in Wallace or Raccoon missed a soap that day, maybe they thought it was sunspots.

■ ■ ■

Jan Roach, the safety director, paced the furnace deck, her eyes traveling over the men. During a lull she said to me, "The first time I saw one of these furnaces, it scared the tar out of me."

Now hot metal attracted her. She liked to stare into arc furnaces, to watch a heat of liquid steel bubbling inside the cave. "My goal is to do every job in a steel mill, to understand how these guys work," she said. "I want to work on a furnace deck."

The arc furnace crackled. She raised her voice. "If you hear the men say 'go,' that really means go," she said. "Don't wait to find out what's wrong. Whatever it is, you don't want to know. Just *go*. They could knock you down on their way out."

At 10:40 A.M., three hours into the heat, Keith Busse showed up on the furnace deck. He wore a spark-proof green jacket. He went over to Mark Millett and asked him how things were going. As they were talking, a crack appeared in the wall of the furnace.

A spear of flame shot out of the crack. The crowd backed away. I stood up, listening for the word *go*. There were several escape routes, and one was a leap of three stories to the ground from the furnace deck. When you are near a ninety-megawatt arc furnace that is coming unglued, the best way off the furnace deck may be a free fall.

Busse and Millett remained calm. They approached the hole in the furnace. They studied the flames, leaning forward. The electrodes were withdrawn and the roof was removed from the furnace. Millett circled the furnace, looking for other leaks.

■

"It's all right," Busse remarked. "It's typical to see little leaks during the first heat. The slag inside the furnace will eventually seal them."

I sat down on my bag of chromium sand, sweating like a pig.

■ ■ ■

The furnace's exhaust duct overheated. Millett ordered a water hose to be brought up and welded to the duct, a quick fix. The welding job took an hour. Meanwhile the furnace was idled, and I went to lunch.

The Crawfordsville Project had a cafeteria that offered abundant food at fifties prices, a dinner for a dollar and a half. They call it mill food. Mill food is one of the glories of a steel mill, featuring brown gravy puddled in fresh mashed potatoes and parti-colored succotash mixed with brown gravy and biscuits with cream gravy and iceberg lettuce with Russian dressing and baked chicken paprikash or New York strip or roast beef and egg noodles with brown gravy and blueberry pie with vanilla-chocolate-swirl ice cream. Steelworkers think high cholesterol is a church holiday.

The furnace men are subdued but cocky, waiting in line for their mill food. One of them says to a cafeteria lady, "Lemme have some of that chicken-fried steak, ma'am. I'm trying to put on weight. You ladies cook too good." She thanks him for the compliment and puts a piece of deep-fried steer on his plate, picking it up with a plastic glove.

Back at the arc furnace, the contents of a second clamshell bucket, sixty tons of Rust Belt, were dropped into the opened furnace. When the cold scrap hit the melt, it drove a volcano of sparks to the roof of the building, a hundred feet in the air, and the sparks rained into the crowd. People cheered. Mark Millett made a fist and shouted, "Yeah!"

"Ain't this God damned something?" shouted Busse to Millett.

"It's going extremely smoothly, considering," replied Millett.

"I just want to keep ramming those electrodes home!" roared Busse, slapping his fist on his palm, and they both laughed.

The truck and bus bundles, the cubes of scrap, began to collapse against the electrodes and short them out. The truck and bus bundles sizzled and broiled and smoked and burst into flames like pieces of suet thrown into a bonfire, and the uproar became awe-inspiring, even heard through ear-plugs.

The furnace began to give off the noise of an artillery bombardment. Thuds and booms and crackles rippled the deck. The air around the furnace deck tingled with a reek of vaporized steel and cooking limestone slag and hot firebrick. It smelled like a skunk.

■

Spoke a chorus of furnace men:

"Now this place really smells like a steel mill."

"If that transformer blows, everyone on this deck will be history."

"You ought to hear it when railroad wheels hit the electrodes. It jars your haid."

A furnace man leaned toward my ear, while his hand figured the air with a bulky glove. "That sound," he screamed. "It dumps your guts. Specially if you've got a lot of guts to dump." He patted his belly and walked away.

■ ■ ■

Millett ordered the electrodes to be withdrawn from the furnace. Silence fell over the deck, except for the hiss of water inside the furnace's cooling pipes. He ordered the slag door to be opened. He wanted to see how things were going inside the furnace. The crowd approached the furnace, kneeling on the deck to peer into the door. A couple of truck and bus bundles had piled up against the slag door and then melted together, blocking the door with a solid wall of frozen steel.

A furnace man picked up a length of steel pipe, an oxygen lance. He fitted the lance into a lance holder—the socket of a high-pressure oxygen hose. He carried the lance across the deck and jammed it into the steel that had clogged the door. He walked back and forth, working the lance around inside the lump of junk, blowing oxygen into the metal. He worked it for half an hour, sweat pouring down his face, until the frozen steel caught fire and burned like pitch and slipped away. He pulled out the lance and then drove it obliquely into the furnace, into a browned gob of metal as big as a piano. The gob burst into flames and fell apart and tumbled into the heat with a splash.

The slag door was lowered, the electrodes descended into the furnace, and the meltdown continued.

Eventually all of the lumps of steel had melted. The furnace contained a bath of live fluid steel. The next thing to do was to raise the temperature of the bath to 2,900 degrees Fahrenheit, and to adjust the amount of dissolved carbon in the bath.

A furnace man thrust a disposable cardboard lance into the bath. The lance burst into flames and dissolved while it sent data to a computer, and a temperature readout appeared in red letters on a board over the control pulpit: 2,822 degrees Fahrenheit. The bath was slightly cool.

To warm a bath of steel, you zap it with arcs. The electrodes were

■

lowered into the furnace until they hovered 2 inches above the bath. The power setting was reduced to 55 million watts, and the arcs were turned on again. A crackling hum came out of the furnace, not as loud as during the meltdown, but loud enough to damage one's hearing without earplugs. The hum continued for a few minutes, and then Millett ordered the door to be raised while the electrodes were cooking the steel, so that he could inspect the arcs as they struck the metal.

The door moved up, and a knife blade of sapphire light streamed out of the furnace—a fierce blue light, more dangerous than sunlight—and it fell on the crowd. The crowd moved backward, people lowering their dark glasses and shielding their eyes with their hands. The arc light whitened the faces of the furnace men, who stared into the furnace. The door opened wide. The crowd pushed closer to the slag door, kneeling and gazing over one another's shoulders, to see what was going on inside the furnace.

I lowered my furnace glasses and looked. I could see blue plasma ropes stabbing from the tips of the electrodes into the bath, cutting into liquid steel. The arcs chattered around the tips of the electrodes and spidered across the surface of the bath. The steelworkers didn't like the spidering.

"That's called 'arcing off,' " said one. "It'll eat up the walls of the furnace."

The temperature of the plasma in the arcs alternated in pulses, from 6,000 degrees to 16,000 degrees Fahrenheit—the boiling points, respectively, of steel and graphite. At a maximum pulse, an arc attained one-and-a-half times the surface temperature of the sun.

The dangerous light of the arcs contained alien tints of aquamarine and topaz. The arc light came out of the slag door in a square blue beam, washed over the crowd, and fell across the wide tarpaulin that hung in the open bay by the furnace deck, wavering on the tarpaulin and on the wall, like reflections from a swimming pool. The tarpaulin billowed in the wind, opening to a view of meadows and distant trees. It was late afternoon. The morning drizzle had stopped, and the air had turned clear and lovely. A rack of storm clouds departed toward the south. The arc light splashed around the tarpaulin and went out into the countryside in pale violet rods. Surely in Whitesville they heard the arcs. God knows what they thought in Whitesville if they saw the light coming out of the mill.

■ ■ ■

After a long-arc meltdown, when a bath of liquid steel is sitting inside an arc furnace, the steelworkers perform a series of operations on the steel.

■

They adjust the amount of carbon dissolved in the steel and they raise the temperature of the liquid steel. This is known as finishing the heat. Millett wanted to produce a finished heat of moderately low-carbon steel, a mild steel, easy to cast.

One step in finishing a heat of carbon steel inside an arc furnace is known as lancing down the heat. Lancing a heat, like other aspects of the craft of making steel, is at least as much a black art as a science, requiring more by way of instinct and experience than book learning. There is no school for furnace workers except the river of steel, and no one knows the river except an older furnace worker, a melter with years of experience. The craft is long to learn. Iverson says that it takes a melter at least three years to learn how to make a good heat of steel in an electric arc furnace. A skilled furnace worker watches the color, texture, and moisture content of the junk that is dropped into the furnace. He watches the length and pattern of the arcs as they play over the bath. He listens to the sound of the arcs. He sniffs the air, wondering if the slag has the right smell. He watches the way the liquid metal bubbles and burps, and he examines the color, thickness, and texture of the slag floating on the steel. Hot metal men have been making heats of liquid metal since at least 3500 B.C., when they smelted copper in furnaces in the Sinai desert, in the days when the book of Genesis tells us that there were giants on the earth. In those days they blew air into furnaces by stepping on goatskin bags. In electric steel minimills they inject pressurized oxygen into a furnace by means of a hollow lance. A good melter develops a personal style for lancing down a heat of steel. Lancing has a formal yet individual quality, like a dance.

A furnace foreman by the name of James E. McCaskill began the task of lancing down the heat. McCaskill was a thin man in his early forties, with a beak nose, a mustache, and deep-set blue eyes, an experienced melter who had worked for Nucor for twenty years. McCaskill had been at the Darlington startup in the summer of 1969, when Iverson's steelworkers ran into the bean fields.

McCaskill picked up an oxygen lance, a steel pipe eighteen feet long, and fitted it into the socket of a high-pressure oxygen hose. There was a hiss and the hose twitched with pressure. The hose contained oxygen at 150 pounds per square inch. McCaskill approached the furnace, dragging the hose behind him. He gave a hand signal. The electrodes were backed out of the furnace and their white tips appeared. A hush fell over the furnace deck, broken only by the hiss of water circulating inside the furnace's walls. Forty people had gathered on the deck to watch the lancing operation. They kept their eyes on McCaskill. An air of expectancy quieted the crowd.

The slag door opened, throwing white-orange blackbody light over Mc-Caskill. He stood in a spotlight. He hefted the oxygen lance in his right hand, sliding his hand along the lance until he found the point of balance. He turned his body sideways and pointed the lance into the light. His right shoulder and hand, holding the lance, faced the steel, while he kept the left side of his body turned away from the steel, with his heart shielded by his body. It is a good idea to keep your heart behind your body when you are lancing steel. Things can fly out of a furnace when you are lancing down a heat. You wouldn't want to take a bit of scrap through the heart.

McCaskill watched the bath, holding the lance at the level of his waist. He might have been a whaleman in a longboat nearing the monster, hefting the lance for the thrust. For a moment time slowed. Then McCaskill raised the lance, profiled the steel, and fell forward and harpooned the whale. The bath rumbled. A blow-back spewed out of the slag door, a firestorm of whirling burning steel sparks and glowing slag grit. The blow-back submerged McCaskill in a fire-cloud. He almost disappeared.

At the moment of the thrust he snapped his body fully sideways and tucked his left hand behind his back, to get it out of the way of the blow-back. With his right hand, he raised the haft of the lance high above his head. That kept the lance tip pointed downward and deep in the metal. He worked the lance tip around in the pool, as if he were feeling for an aorta. The steel protested with shrieks, blow-backs, rumbles, coiling movements, and an unbearably bright light.

The metal boiled: a carbon boil. The oxygen combined with carbon in the steel with a thunderclap, and the metal frothed. Sparks hissed out of the furnace, rained over McCaskill's green suit, ricocheted off his hard hat and chest and dark glasses, bounced and skittered across the furnace deck and sailed clear into the air, outdoors, off the edge of the deck. With a great beating of flukes the carbon boil intensified. McCaskill dodged his left hand up to his face to protect it from sparks, peering into the slag door over his glove. The whites of McCaskill's eyes flickered at the corners of his dark glasses; he was scanning the bath for signs of trouble. He worked the lance left and right, never taking his eyes off the heat, never turning his back to the steel. No furnace man ever turns his back to hot metal that's within spitting distance. He moved closer to the door, because the lance was melting away.

During a quiet moment, later, McCaskill lit a cigarette and recalled

■

Nucor's first heat, on that June day in 1969. "I had never seen a damn steel mill before," he said. "I was up on the casting deck when they started that furnace, but I seem to recall there was a lot of runnin' going on. I was all ready to run myself, if I'd have known where to run to. A man's got to be crazy as hell if he ain't a little jittery around one of these furnaces during a startup." Twenty years later, McCaskill had accomplished some 35,000 heats of steel. "Lancing?" he said. "That's nothin' strenuous."

Lancing melts the lance. McCaskill withdrew a glowing stump of lance from the bath. The tip of the stump was coated with steel and glowed with a hurtful light. He wrapped a gloved hand around the stump and unlocked the stump from its socket. The stump flopped on the floor, a glowing soft noodle of pipe. He handed the oxygen hose to another melter. This man fitted a fresh lance into the socket. He stood before the slag door, profiled the steel, and drove the lance into the beast. He lanced for a little while and then he passed the lance to another melter, who put the lance into the steel. The melters shared the honor and glory of lancing the first heat of steel.

■ ■ ■

A carbon boil, triggered by a lancing operation, raises the temperature of the bath until the steel is runny enough to be poured like water into a casting machine. Temperature readings and samples of the steel are taken after a lancing operation. Lancing down steel also helps to bring a slag up to the surface of the bath. As the slag rises, it brings impurities with it. Lancing cleanses steel.

A steelworker thrust a steel rod into the bath. On the end of the rod there was a cup. He fished the cup through the bath and withdrew the rod. The cup was now yellow. The cup hissed as he plunged it into a bucket of water. He withdrew it from the water and knocked the cup on the deck. A black hemisphere of steel fell out of the cup. Jeri Switzer, the chemist, carried the sample into a chemistry lab at the metallurgical deck. A few minutes later she came back and reported that the carbon content looked about right.

A steelworker slid a temperature probe into the bath. The bath had been raised to a temperature of 2,965 degrees Fahrenheit.

Mark Millett knelt before the open slag door and examined the heat through a piece of smoked glass, to look at the slag that had risen on the bath. Keith Busse knelt next to him. The electrodes descended to within two inches of the bath. There was a spitting sound, and a haze of sparks

■

whirled from the bath. Millett and Busse jumped backward, and arc light threw their veering shadows on the blue tarpaulin. Then they both moved forward again and studied the electrode tips. Fire gnats buzzed around their heads. The bath often popped with small explosions and rushes of sparks. The sound of the arcs was a high snapping crackle. Bright flashes came out of the furnace. Then the electrodes were pulled away, leaving a calm, flat bath of liquid metal. It was nearly a finished heat of steel.

I knelt behind Millett and Busse, watching the heat over their shoulders. When the arcs were turned off, the light coming out of the slag door came from the metal itself, from the hot electrodes, and from the hot interior walls of the furnace. Everything inside the furnace glowed. Looked at without furnace glasses, the open door was a bright featureless hole in the side of the furnace, a uniform yellow glare that hurt the eye. I lowered my dark glasses.

Seen through smoked glass, darkly, the metal reflected light. The molten steel had a mirrored surface; it was a fascinating liquid, with a hint of silvery gray, a dull quicksilver, the color of plain carbon steel. But it also glowed with a powerful light that shone from beyond the mirror—the blackbody light of liquid steel. The pool rippled with twitchy movements. Limestone slag bobbled and danced on the heat. The metal simmered, bubbles pushing the slag around like a foam on pea soup.

The men and women on the furnace deck could not take their eyes off the heat. They kept moving toward the furnace door and being driven back by sparks, holding their hands up to break the light from their eyes and faces. Daggerlike flames shot up through the slag. The flames had green shafts and sharp yellow tips. The interior of the furnace recalled a bed of daffodils blowing in the wind.

Keith Busse kept his eyes fixed on the pool. He had seen molten steel before, but this time it was the first heat in his first steel mill.

"Most people wouldn't see this as beautiful," he shouted, turning around on one knee. "It would be awful and ugly to them."

The furnace burped—"*Gunk!*"—throwing sparks over Busse. He glanced at the furnace, getting ready to jump backward. "But most people wouldn't come into a place like this anyway," he said. "The only people who ever come into a melt shop do it for a reason. They're in here to make a living."

The light cast huge shadows of men and women on the wall of the steel mill, of steelworkers carrying oxygen lances, of technicians hurrying back and forth, of managers in hard hats huddled in knots, shouting into each other's ears over the thunder of the furnace. The shadows loomed larger than the people, and yet they seemed to belong naturally to the people. In a steel mill it is not hard to imagine that there are still giants on the earth.

■ ■ ■

At 3:30 in the afternoon, Mark Millett declared a heat of steel. The heat was at a temperature of 2,900 degrees Fahrenheit, and it was a mild steel, without much carbon in it. Millett gave orders to prepare to tap the furnace, to drain the furnace into a ladle.

A ladle crane began to creak along rails at the roof of the melt shop. The rails ran near the tops of Duane Hurler's columns. The crane moved the length of the melt shop to the ladle preheat station, the low deck near the base of the casting tower where gas burners had been cooking two ladles since eleven o'clock the previous night. The crane picked up a ladle. The ladle was a tapered bucket, 15 feet tall. It could hold 120 tons of liquid steel. It was lined with bricks, now glowing red-hot. Heat waves shimmered from the mouth of the ladle. The crane carried it back up the length of the melt shop, lowered it, and placed it on a ladle car. The ladle car resembled a railroad car. Bells rang, and the ladle car moved the ladle underneath the arc furnace.

There was a hole in the bottom of the furnace, the tap hole, for draining steel out of the furnace. The tap hole was covered by a sliding door. With the ladle now stationed below the furnace, the furnace was tilted—it rested on rockers, like a rocking chair—and an air cylinder stroked open the door under the tap hole.

A thread of sand fell out of the tap hole. That was all. Not a dribble of steel came out of the furnace.

Two furnace foremen ran down a flight of stairs to an observation platform below the furnace, to try to look up inside the tap hole to see what was going on in there. They leaned over a railing and craned their necks, and then returned up the stairs and reported to Mark Millett that the furnace's tap hole was plugged with a skull, a lump of frozen steel. The hole was welded shut with steel.

"We're going to have to lance that son of a bitch out," said a foreman.

To lance a clogged tap hole is a tricky procedure. Hot metal men blow oxygen into the hole with a lance. The solid plug, or skull, catches fire when bathed in pressurized oxygen. The skull burns like a piece of tar and dribbles out of the hole, opening the hole. The furnace's tap hole usually opens suddenly, releasing a gusher of hot metal. You must stand below a furnace to lance it, but you don't want to stand too close to the hole.

■

Mark Millett, James McCaskill, and two other furnace men decended the stairs to the observation platform, carrying lengths of steel pipe, lance pipe. They fitted the pipes together, fashioning an extra-long lance. They fitted an oxygen hose to the lance. All four men held the lance steady, and they fished the lance up into the hole in the belly of the furnace.

I stood with a crowd on a walkway near the tap hole, watching Millett and his men poke the lance into the furnace. They turned on the oxygen. There was a hissing sound. They probed the lance delicately into the furnace's underbelly. Suddenly there was a brilliant flash and a deafening explosion under the furnace. Something had gone wrong.

The furnace men went up the stairs in a catlike blur, so fast one hardly saw the movement, clawing at each other, climbing each other's backs, with their mouths clamped shut and their eyes white with panic. Millett flailed at the center of the knot. They left their harpoon behind them, suspended in midair, and it fell into the depths of the melt shop, trailing its loose hose.

When professionals are running, it's time to go. The crowd was buffaloed, breaking northward along the furnace deck, but Millett and the furnace men gathered at the top of the stairs and regrouped, peering at the furnace, while their words rolled over the deck: Fucking hell! God damn it! They started to laugh: That son of a bitch popped. Nothin' but a damn oxygen pop.

The oxygen gas had entered the hole and found iron gas in there. The two gases mixed and exploded with a crack like a stick of dynamite.

This pop scared them a little bit. Later a furnace man remarked, "It was all asses and elbows after that explosion. And I tell you what: James McCaskill was the first man up the stairs." Thirty-five thousand heats and still a little jittery, that James McCaskill.

■ ■ ■

The tap hole was not damaged. Millett and the melters returned down the stairs and lanced the tap hole a second time. There was a second flash and boom. The pop routed them cursing up the stairs. They regrouped for a third try, and this time, as they probed the lance into the belly of the furnace, there came a *whoosh*, a burst of warmth, a sound like the wind, the furnace let go, and the pour began.

A murmur went up from the crowd. The murmur rose into rebel yells as the pour intensified. The furnace men scattered up the stairs through a turmoil of sparks, waving their fists. They had tapped a heat out of a

■

furnace, one of the oldest rituals in human industry. A splendid golden light filled the melt shop, coming from the underbelly of the furnace. The furnace had given birth.

A hundred and twenty tons of molten steel drained into the ladle, a process that took three minutes. The pour ended suddenly with a drizzle. The furnace men heaved bags of limestone into the engorged ladle. The bags hit the steel and burst, spilling powdery rock on top of the molten steel. The rock congealed into a white slag blanket, to lock warmth into the metal and to prevent air from touching it.

The traveling crane picked up the ladle, carried it to the metallurgy deck, and placed it on a car. The car moved the ladle beneath a triad of electrodes. The electrodes were plunged through the white slag into the steel, to toast the baby with 18 million watts of electrical power, to keep it warm, while chemists ran tests on it. They reported that it was a clean steel.

Now it was time to kill the steel. The killing mechanism was a wheel that unrolled a length of aluminum wire into the steel. As the aluminum wire entered the steel it dissolved, mixed with the steel, and removed the oxygen from the steel, and so the steel was killed.

Since the casting machine wasn't ready, there was nothing to do but to pour the killed steel onto the floor of the melt shop. The ladle crane lowered the ladle down to the floor and tipped it over, and steel dribbled into a line of pits dug in the floor. The steel hit the dirt with a frying sound. The skull pits filled up, one by one, like water going into an ice tray, as the ladle moved over them, flooding the west end of the melt shop with a red glow and an ovenlike dry heat. The heat pushed the crowd backward on the metallurgy deck, people shielding their faces with their hands.

There was a loud spitting noise. The crowd moved farther back. One of the skull pits on the floor erupted like a volcano. The pit contained a little bit of damp mud, and water in the mud flashed to steam, and the mud exploded and kept on exploding beneath the puddle of liquid steel. The noise was remarkable, and flames leaped up from the pit and detached themselves from the ground, screwing upward through the melt shop in towering salamanders. Golf balls of liquid steel were chipped twenty feet through the air. You can put water on steel, but you can't put steel on water.

Keith Busse lowered his dark glasses and grinned. "I never saw Mark so frightened in his life, scrambling up those stairs! I called Mr. Iverson. I told him we'd tapped a heat in ten hours. What a marvelous day!"

■

The chorus spoke in many voices:

"That went pretty smooth, like a cow on ice."

"I tell you, it was a long stir gettin' here."

"As of now, this isn't a construction project anymore, this is a steel mill."

"I saw liquid steel for the first time in my life today. It looked like Mars. The power that stuff puts out. I planned my escape."

"I was as nervous as a dog shitting on a log chain."

"This steel mill is going to be the making of Crawfordsville."

"One thing is for sure: the first cast won't take any ten hours. Once they put steel into that caster, it's like lighting a fuse on a stick of dynamite. You light a fuse and something's going to happen real quick."

"A couple of months from now they'll be making heats every day in here and nobody will care."

"Twenty years from now you'll be just another bald guy running a machine."

"I'll tell you a little speech Iverson give us. He said, 'I know what the American Dream is. The American Dream is to make more money this year than you did last year.' And I tell you what: I got sixteen hundred and fifty shares of Nucor stock, and as of this morning it was worth ninety thousand dollars."

"We all have a dream, right?" said a woman in a hard hat, a grandmother by the name of Earleen Kurtz. "I farmed eighty acres. I raised corn and soybeans, a little bit of wheat. I raised a few hogs, fifty head was all. You can't farm eighty acres. I sold the land and got out. I want to be a ladle crane operator. I want to carry steel from the furnace to the caster. I don't know if they'll give me that job. They only want the most experienced people to carry ladles of steel. I'll probably end up on a scrap crane. But to carry a ladle to the caster, that's the dream. That's the dream." Such were the words of Earleen Kurtz, grandmother and steelworker.

A ladle crane operator sat on a bag of metal sand, watching the steel fire erupting on the floor. His name was Mike Hinz, and he told me his dream. Hinz was a quiet young guy with a brown beard, a round, rough face, and a rather shy laugh. "I'm hoping my retirement plan will hit six figures," he said, raising his voice over the uproar of the steel fires. "Then I'm going to buy my mobile home and retire."

■

14

DRY CAST

■ ▪ ■

Janice Roach, the safety director, prowled the casting deck. She stopped a group of steelworkers. "Hey! There's a no-jewelry rule up here," she said. The steelworkers wore gold neck chains and gold bracelets. "You guys have to get all that jewelry off."

They didn't know what to make of her safety rule. "All of it, Jan?" said one man, pointing to his wedding band.

"Yes," she said. "There's a danger of solvents up here, and you guys wouldn't want to damage your wedding rings. And your jewelry could catch on things."

She was five feet three inches tall. Her eyes flickered behind her glasses, looking left and right and up and down as she poked into corners and inspected trash cans. She wore a green spark-proof jacket, and her hard hat displayed a red cross. She stopped a steelworker and pointed to a wheelbarrow full of empty Coca-Cola cups. "I want this out of here before we put any heat on this deck."

He nodded.

She continued her patrol. "There's a lot of tension up here on the deck today," she said to me. "These guys seem real tense. I don't like that. That kind of tension falls one step beyond respect for steel into fear of steel. If the tension remains this high, we won't cast. The general line is that we won't start the machine until it's ready, but are you ever ready?"

"How do you feel?" I said.

"How do I feel. I don't know how I feel. I feel nervous. During a time

■

like this you eat steel, you sleep steel, you think—'' She beckoned to Big Dave Thompson, casting supervisor. ''Dave—get these nitrogen tanks out of here. Will you do that for me?''

Thompson, chewing a toothpick, said he'd oblige.

Janice Roach had been through Nucor startups before, but this was her first startup of a steel mill. ''This doesn't seem to be a normal startup tension,'' she remarked to me. ''There's something more than that going on right now. Keith Busse is pacing in his office. We always know when he's pacing. You can hear his boots in the hall, going in and out of his office. He's getting himself half cups of coffee all day.'' She looked at me with a penetrating expression on her face. ''Do you know where the exits are up here?'' she said.

Two stairways led down from the casting deck, at the east and west sides of the deck. I pointed to both.

''Good. Know where they are.''

The exit stairs descended thirty-two feet to the floor of the mill, in flights interrupted by landings. ''The men have timed themselves. They can get down those stairs in seventeen seconds,'' she said.

She warned me to listen for the word *Go*. ''If you see the men start running, you start running too,'' she said. ''They'll converge on the exit stairs. They'll probably knock you down on their way out. Go over the rail if you have to—having a broken leg is better than having your feet melted off. Keep in mind that steel can't run uphill. You want to get above the steel if you can.''

Liquid steel weighs 484 pounds per cubic foot. It wants to go downhill in a hurry. Steel is eight times heavier than water. When the metal is in a fluid state it flows with remarkable speed. A heavy spill can cover a casting deck in the blink of an eye, and if you stand ankle deep in a big splash you will probably die, because the steel gives off enough radiant heat to burn the skin off your body.

''What's the plan of action, in case of an accident up here?'' I asked.

''The casting supervisors have talked about this. They've had many discussions,'' she said. ''They'll tell their men to abandon ship, but they'll stay with it.''

So that was the idea. If the casting machine came unglued, Big Dave Thompson would be on deck fighting steel while the steelworkers ran for their lives.

''How come I don't see more women around here?'' I asked.

A welding torch popped, and we turned away from the glare. ''Most

■

193

women are not confident around these mills," she said. "It's not that they can't do the job or that they aren't smart enough. But I've seen administration girls who can't put one foot forward when they get inside a steel mill. I know how they feel. The first time I was in a steel mill, I was scared to death. I said to myself, 'You know, what is wrong with me?' There was nothing wrong with me."

Jan Roach believed that the men were nervous. The men hired locally had had six weeks of training at Nucor steel mills or in West Germany at the SMS foundry. They were afraid, but they wouldn't admit their fear to each other and certainly not to their boss, Mark Millett. But when they talked to Jan Roach in private they revealed to her what she called their "inside feelings." "They say to me, 'We don't really know what's going on here, but hopefully everything will be O.K. But we'll just have to wait and see.' I think they're afraid that if it goes, it'll go in a massive way— everything is so large up here. It's the simple worry that things won't hold together."

She noticed a drum of slimy liquid covered by a couple of loose boards, sitting on a landing on the exit stairs. Motioning to Dave Thompson, she said, "That barrel and those boards are blocking the staircase."

"I'll take care of that now," said Thompson. Thompson tossed the boards over the rail. They sailed to the floor of the steel mill.

"What about this barrel? I want this removed, too."

"It's just water."

She peered into the barrel. "That's not just water, Dave, there are bricks in there. You won't be able to move this barrel if you need to." She dipped her finger into the liquid, stirred it around, and pulled up a coating of brown slime on the end of her finger. She put her finger in her mouth and tasted the slime. "That's oil," she said. "Get it out of here."

"I'll throw it over the rail," said Thompson.

"Good."

Thompson dumped the barrel, and greasy water cascaded off the casting tower. Then he dragged the barrel, half full of bricks, down the stairs and out of the way.

In a steel mill, any miscalculation of natural forces or any clumsiness of human art could trigger a fireball. Anything could go wrong during a startup. The task was to try to imagine the possible ways in which a complicated system might fail, especially if the failure could tear someone apart or coat him with metal. The other task was to keep warning the steelworkers in a loud voice to never stop thinking around steel.

■

A tunnel ran beneath the casting tower, and Jan Roach wanted to inspect it. I followed her down a flight of stairs below ground level. We walked the length of the tunnel, splashing through puddles of water, until we entered a concrete chamber lit by a naked light bulb. A plastic pipe was embedded in the concrete. "Where's that pipe go?" she muttered. "It's plastic. If steel got down in here the plastic could burn."

Back at ground level, she found a bucket full of black goo. "Who left this here?" She put her finger into the goo and brought up a shiny glob. She brought it close to her nose and smelled it.

"Are you going to put that in your mouth?" I asked.

She smiled. "It's the easiest way to find out what something is—you taste it. That's quicker than sending it to a lab." She wiped the black teardrop on a steel column. "I didn't taste that one," she said. "The guys are telling me that someday I'll burn my tongue off."

■ ■ ■

Steel would drop in two stages into the casting machine, pulled downward by the force of gravity. In the first stage, steel would drain from a hole in the bottom of a ladle into a tundish, a bathtub, an intermediate holding tank. In the second stage, steel would drain from the bathtub into the machine. Steel went from ladle to bathtub and from bathtub to machine; two stages, under gravity, like water flowing through locks.

A tundish operates exactly like a bathtub. It has a drain hole in the same location as the stopper hole in a bathtub. The drain hole is plugged by the stopper rod, and it leads to the injection nozzle, a white flaring ceramic tube that hangs from the belly of the bathtub, protruding downward between the lips of the Kolakowski mold.

The hole in the bottom of a ladle, through which steel drains into the bathtub, is opened and closed by means of a sophisticated valve, the ladle slidegate. On the floor of the melt shop, not far from the casting tower, there was a small platform where steelworkers repaired ladles and heated them with gas flames. Mark Millett took me to the platform to show me a ladle slidegate. A ladle sat on the platform, tipped over on its side, and a slidegate had been fitted into the bottom of the ladle. The slidegate was a black tube, with a two-inch bore diameter, that stuck out of the bottom of the ladle. It looked like the mouth of a small cannon, and it was fitted with a complicated assembly of moving parts. Millett pulled a pencil out of his pocket and stuck it into the slidegate and circled the pencil around.

■

"The ladle slidegate is a critical part of a steelmaking operation," he said. "If the slidegate fucks up—I mean if it malperforms—we'd be in trouble."

Millett peered through the slidegate hole—looking up into the huge ladle through the tiny hole—and then slid his finger around inside the hole. "If slidegate parts were made of steel, they wouldn't survive very long," he said. "Everything that touches the steel stream in the slidegate is made of alumina-graphite."

Alumina-graphite is a black compound made from graphite mixed with powdered synthetic sapphire. At maximum flow, the slidegate could pass one ton of steel every eighteen seconds. Through that two-inch hole a million tons of steel a year had to flow. A ladle slidegate is the barrel of an automatic weapon. Pressurized steel flows through the hole at a pressure of as much as fifty pounds per square inch. The pressure comes from a twelve-foot head of steel sitting in the ladle. The steel also heats the slidegate to a quarter of the surface temperature of the sun. Not even sapphire can last long under those conditions. Ladle slidegates deteriorate quickly, and a slidegate has a way of locking open during a pour. If it can't be closed, it can dump steel all over the place.

The sapphire slidegates were made by the North American Refractories Company. A North American specialist worked at the small platform, at the base of the tower, along with a Nucor crew, servicing slidegates. The North American specialist was a certain Patrick Major. Millett often plucked his walkie-talkie out of his hip pocket and said, "Pat Major! Pat Major!"

■ ■ ■

Ladle slidegates frequently become jammed with a lump of steel, a skull. In 1974, at Nucor's Darlington minimill, the slidegate of a full ladle of steel sitting at the top of the casting tower became clogged with a skull. Steelworkers used an oxygen lance to burn open the slidegate. Somehow they managed to bubble a great deal of oxygen up through the slidegate into the molten steel in the ladle. The oxygen mixed with the steel. Now it was live steel, fizzing with oxygen. The ladle erupted in a carbon boil and foamed over like a beer mug. Four steelworkers were caught in their tracks and killed. They shriveled like ants under blazing gasoline. It was the worst accident in the history of the Nucor Corporation.

■ ■ ■

■

Joseph Kinney, the executive director of the National Safe Workplace Institute, in Chicago, does not love the Nucor Corporation. He trails around after Nucor, visiting small towns where Nucor operates, talking to people and gathering reports on accidents at Nucor plants. Kinney describes his group as "the only national organization whose sole focus is on employee safety and health." I called Kinney on the telephone to hear what he had to say about Nucor. He had plenty to say.

"Frankly, from our point of view, Nucor represents what is wrong with American business," said Kinney. "At Nucor the burden of safety is on the plant manager. My belief is that when you operate in a decentralized environment, without other kinds of corporate checks, you can have health and safety problems. The company has a horrifically high fatality and injury rate. Nucor has at least quadruple the fatality rate of the steel industry as a whole. The steel industry's fatality rate is just about on the average for American industry, at about 8 deaths per 100,000 workers per year. I don't have exact figures on Nucor, but Nucor's fatality rate looks to be 50 or 60 deaths per 100,000."

Kinney's figures appear to be exaggerated. If Kinney's claim is accurate, then Nucor is losing shocking numbers of people in operational fatalities, two or three accidental deaths a year, making Nucor's factories regular hellholes, in Kinney's opinion. But to the best that I can discover, Nucor seems to be losing about *one* employee a year in accidental deaths. James Coblin, Nucor's personnel manager, provided me with what he said is a complete list of Nucor's accidental fatalities between 1980 and 1990. The list describes 9 deaths in 10 years—3 deaths in Nucor's steel mills, 6 in the Vulcraft plants—out of a work force that averaged 4,500 people during that time. It works out to 20 deaths per 100,000. That's 2 ½ times the national average. But not 6 or 7 times the national average, as Kinney claims.

"Years will go by without any fatalities," Iverson said to me.

"There's just no way to know if Nucor is telling you the truth," Kinney said to me over the telephone. But Kinney himself can't produce any figures to back up his claims.

Kinney is starting to drive Nucor's management crazy, and Nucor, in turn, seems to be driving Kinney crazy. Kinney thinks that Nucor's bonus systems and the absence of work rules encourage both managers and workers to take chances with safety in order to increase production.

Coblin, the personnel manager, disputed that charge. "Kinney," he said, "just feels that all production incentives are against safety. But we

■

197

have found that when we are the busiest, when we are cranking out the tons, we get the highest safety. We have accidents when we're running slow, when we're starting and stopping a plant a lot.''

That is what happens in a startup. The plant starts and stops frequently, as machines go haywire, as the process comes to a halt and then has to be restarted. The fact is that Nucor builds factories fast, starts them up fast, and staffs them with young people who have little or no experience in manufacturing operations. Young, inexperienced workers are killed more often than seasoned veterans, and this is particularly true when the workers are trying new production methods and starting up new or unusual equipment. Furthermore, because Nucor has virtually no corporate staff—not a whole lot of employees working safely behind desks—most of the employees at the company come into daily contact with machinery. It is not surprising that Nucor has a fatality rate that is more than twice the national average.

Or is Nucor's death rate actually closer to the average than it would seem at first glance? More than half of Nucor's labor force works in Vulcraft plants, making joists and decking. According to the Bureau of Labor Statistics, plants that fabricate structural metal products are *twice* as dangerous as primary steel mills that make raw steel. Nucor's death list confirms this statistic. In ten years at Nucor, twice as many people died in the Vulcraft plants (six deaths) as did in the steel minimills (three deaths). Now Nucor's death rate looks a bit more ''normal.'' The other question, of course, is whether the death rate is tolerable.

■ ■ ■

Big Dave Thompson, bearer of a master's degree in safety administration, ran his men through drills to make sure that they were comfortable with moving a ladle around the casting deck, to open the locks and let the river of steel run into the machine. He called it sequencing. ''We want to get good sequencing here,'' he said to his crew. ''Before the pour, you move the ladle car to here.'' He stood on a chosen spot. ''We should mark that,'' he said.

A steelworker marked the deck with a can of spray paint.

Pushing buttons and levers on control panels, the crew moved an empty ladle, sitting on a ladle car, across the deck on high tracks, to the clang of warning bells. A black pipe of sinister appearance, the nozzle of the slidegate, protruded downward from the bottom of the ladle. The ladle halted over a bathtub. The bathtub was suspended over the mouth of the

■

casting machine on its own set of rails. That set up the conditions for a two-stage metal-fall: ladle to bathtub, bathtub to caster.

"Open slide!" said Thompson.

A steelworker named Les Pelfrey, the slidegate operator, threw a hydraulic lever and the slidegate moved sideways, opening a hole in the ladle.

"If the slidegate opens right away, no problem," said Thompson to his men. "If it won't open, then you've got to burn it open."

Thompson pushed an oxygen lance up into the nozzle of the slidegate over his head, standing to one side of the slidegate. (You never stood directly below a slidegate, because a slidegate was as dangerous as the muzzle of a gun.) "You guys learned how to burn open a ladle in Nebraska," he said, fishing the lance straight up into the bore of the slidegate. "You want to burn straight up through the nozzle bore, so you won't damage the slidegate. Why don't you go ahead and practice all that?"

They dodged in and away from the ladle, stabbing the slidegate with fine lances.

Pelfrey, the slidegate operator, wore ovoid sunglasses and had a deliberate way of speaking. He suddenly turned to me and remarked in a low voice, "Steel mills, they're hot, dirty sons of bitches. But I never seen such easy money in my life. Unless something goes wrong." Pelfrey had worked at the R. R. Donnelley & Sons printing plant in Crawfordsville, which manufactured books. When a book machine cracked up, sheets of paper jammed the rolls, and that did not entertain Pelfrey.

■ ■ ■

A startup in the steel industry attracts a large number of gawkers, bit-parters, and coffee-cuppers, like the filming of a chase scene on the streets of Los Angeles. The casting deck became a place of feverish activity. Changing groups of SMS engineers, Nucor steelworkers, contract engineers, vendors, hangers-on, curious third parties, and perhaps paid industrial spies gathered in clumps, talking and gossiping. Crowds gathered under an empty ladle, which sat on a ladle car over the deck. The ladle threw a deep shadow over the people and over the casting deck.

Mark Millett roamed the deck, stopping here and there to chat with steelworkers and engineers, slapping someone on the shoulder—"Super!—Excellent!—Ahoy, mate!—What time is it?" To his walkie-talkie he'd say, "Pat Major! Pat Major!"—wanting some advice from the slidegate specialist. He was always removing his hard hat and running

■

his hand through his hair. His large dark eyes flickered restlessly all the time, except at moments when they locked onto something with a terrible intensity.

A sweet odor of exotic lubricants mixed with solvents hung in the air, and over that drifted a nauseating odor of some putrid thing being roasted. That smell came from the tunnel furnace. The tunnel furnace had been fired up with natural-gas jets, and it had been heating for days now. It was lined with a soft, mushy insulation, and as the material became red-hot, it gave off a smell that compared unfavorably to the musk glands of a weasel. "That tunnel furnace smells bad enough to gag a maggot," said a foreman.

The casting machine was governed by computers, and a glassed control pulpit faced the casting deck. Inside the pulpit, the SMS Concast commissioning engineer, Patrick Owens, sat all day and much of the night at a colored video screen. Maroon shadows encircled Owens's eyes, which were red-rimmed and fiery and blue, behind black-wire-rimmed photosensitive eyeglasses. Owens, a slender fellow in his early thirties, smoked Merits around the clock, dropping butts under his video console.

Owens dropped a butt and probed around with his foot to kill it. He gestured out the pulpit window toward crowds of Nucor steelworkers on the deck. "A lot of these guys will work here for the rest of their lives," he said. "They'll never see pressure like this again. I live with this kind of pressure because I commission casting machines. That's what I do: I start up casters."

"I've never seen liquid steel before," remarked a Nucor man.

"Yeah?" said Owens. "In a few days you'll see steel hit the machine, and all hell will break loose."

Owens discovered that his Merit had burned to its filter. He threw it under his feet. He found a full pack of Merits, stripped it, and shook the cigarettes all over the console. About half of them fell to the floor. He picked up a cigarette from the console and placed it on his lip. He ignored the cigarettes on the floor—kicked them aside with his foot. He gathered the rest of the cigarettes into a pile on the console within easy reach. His hands seemed a little shaky. "You don't know if a casting machine will work until you pour steel into it," he said.

Into the pulpit walked Manfred Kolakowski. Kolakowski had flown in from Germany to help start up the casting machine. "How's everything going?" he said to Owens, in a cadenced German accent.

"I have *no* idea," said Owens.

"Don't worry," said Kolakowski. "At SMS we have the technology.

■

We can fix anything. You know the song, 'She changes from day to day'? Our machine, she changes from day to day.''

The inventor hurried across the casting deck to the mouth of the machine. He stared into the slit of the Kolakowski mold. He wore a dapper blue-and-white candy-striped shirt and a yellow hard hat. A pair of elephantine eyes glittered below Kolakowski's hard hat, as Kolakowski surveyed the mold of his imagination. "We want to *ca-a-ast,*" he said, making a fist.

■ ■ ■

Ken Kinsey, the steelworker whom Duane Hurler never got around to murdering, had trained with Kolakowski in Germany, helping to operate the pilot machine. "Manfred Kolakowski's a real down-to-earth guy," remarked Kinsey one day to the casting crew. "He told me one night in a bar, 'You must call me Manfred.' We were having a beer and I had been calling him 'Mr. Kolakowski.' ''

"I don't think he's a mister. I think he's a doctor," said one of the crew.

"He's just a regular guy," said another.

Mark Millett said, "For Kolakowski, this is a dream come true. Under his quiet character, I'd say he's an excited man.''

"Kolakowski's a wizard," said one Dan Harmon, a steelworker. "Any question you ask him, he don't have to go look it up in a book. He's a caster encyclopedia.''

Kinsey said, "I asked him if he thought the casting machine would really work. He said, 'I have a lot of confidence in it—I invented it.' I felt about this high," said Kinsey, showing the crew a small gap between his thumb and index finger.

"Manfred. That's a weird name.''

"Hell, Manfred's a common name over there," said Big Dave Thompson, joining the conversation.

"The thing that kills me," said a foreman named Tom Miller, "is when they go on in their Deutschland or whatever they call it. They go on and on in their Deutschland, and you don't know *what* they are saying. They could be saying anything about you. And they probably are.''

Kinsey said, "They probably think that the minute they get on an airplane and go home to Germany, we'll burn this place down. I heard one of them say, 'And these Americans, they went to the moon?' ''

A steelworker peered down into the lens-shaped opening of the Kolakowski mold. "What did you guys call that, a taco?''

■

"A vagina," said one.

Kolakowski walked over to the men. "Start the machine slowly," he said to them.

One of them said to Kolakowski, "How many casts have you done?"

"We have done five hundred and forty-nine casts at Buschhütten now, and I have missed seeing three of them," replied Kolakowski. He had come to respect the men from Crawfordsville when he had trained them in Germany, but he harbored certain doubts. "These Nucor workers are very like comrades," as he put it. "They are not the sort for sticking around and putting their hands in their pockets, eh? They are very keen. But of course, to once turn a button does not make you a *Profi*, a professional."

■ ■ ■

A steady stream of German executives trickled into Crawfordsville, filling the motels in the town. Franz Küper, the SMS project supervisor, who had been the Americans' handler in Germany, arrived to supervise the SMS efforts during the startup. Küper took one look around the casting deck and went into a fit of nerves over industrial spies.

He buttonholed Mark Millett. "There are guys from the Vesuvius Company walking around here," he said. "What are those Vesuvius guys doing here? You know these guys sell nozzles and refractories to our competitors, to Demag, to Danieli! They are bound to tell our competitors things. They are looking at all the details of the machine. They are checking the motors. I found them checking the machine! Down below there! Where no one is supposed to be! We have an agreement for secrecy with Nucor. It's Nucor's responsibility. It's *your* obligation to tell them to stay away from the machine!"

Millett explained to Küper that the Vesuvius engineers had signed the secrecy agreement.

Küper wasn't pleased about that. He cornered a Vesuvius engineer. "Who let you in here?" he said.

The guy replied, "Hello, Franz! How are you?" He held out his hand.

They shook hands. It turned out that Küper and the Vesuvius man had known each other for years.

"What are you doing here, exactly?" said Küper.

"I'm making field adjustments," said the Vesuvius man, putting a cigarette in his mouth and snapping a Bic on it.

"I don't know who let you in here," said Küper.

"Nucor said it was O.K."

■

"You're just a supplier."

"Yeah, but we've been supplying nozzles to the pilot plant in Germany. We had to sign your confidentiality agreement before we could supply the nozzle. How are you, anyway?"

"I'm fine," said Küper.

"Last time I saw you it was in Granite City, Illinois."

"Oh, *ja!* The Granite City startup!" said Küper, thoughtfully.

"That must've been five years ago."

"It was in 1981."

"You look well, Franz."

"That was seven years ago. My hair is little grayer now, *ja?*"

"You look good, I swear. We meet every five years or so at a startup."

"Then I don't see you for five years," said Küper.

"I guess we'll meet again in five years," said the Vesuvius man.

Later the Vesuvius man explained to me, "I can't say I have a good reason to be here. I just like to be around startups. I can't keep away from these things. You can just feel the tension on this deck. I can't keep away from these things." He dropped a cigarette butt under his foot and snapped his Bic on a fresh bone. "I smoke like a fiend during these things."

In the control pulpit, an engineer remarked to Owens, "There's nothing ike starting a caster. It's exhilarating."

"Is there madness in your family?" said Owens.

"We spend all our weekends up here, and we only get forty bucks extra a day," said the engineer.

"But where else can you get this kind of fun?" said Owens. "Hey!"— turning to someone—"Got any cigarettes?" He accepted a Winston and burned it with his Scripto. He spoke into his walkie-talkie microphone. "All right, let's do it again. Ready? Cast mode."

■ ■ ■

The dry casts began, also known as cold casts, because the machine is run without steel. The casting crew began a dry cast by moving an empty ladle and a bathtub over the machine. A dry cast began with some shouting, and then the ladle slidegate was opened, and imaginary steel began to pour into the bathtub. Then they opened the bathtub's drain plug, letting imaginary steel run through the nozzle into the mouth of the Kolakowski mold.

Thompson had hung a stopwatch around his neck to time the events of a dry cast. Having played football on a college team, he thought of a dry

■

cast as a practice football play. He, Thompson, was the quarterback, Mark Millett was the coach, and Ken Iverson was the owner of the team. Putting together a cast was like putting together a play. It was a matter of moving people through points in space at the right moments in time, slowly at first, then at full speed. When the real moment came, the team would have to perform the cast smoothly, on the beat, hitting the points without hesitation, and he hoped that the crew would resist panic if something went wrong. An accident with liquid steel could paralyze a casting crew, and you might end up with a hundred tons of hot metal where there shouldn't be any hot metal. An uncontrollable heavy spill could burn up a casting tower and could certainly kill people.

"Bring the ladle to cast position," said Thompson.

A one-hundred-and-twenty-ton ladle, fifteen feet high and twelve feet across, creaked along the rails and stopped above the bathtub. The bathtub rested on rails over the mouth of the machine, with its nozzle inserted into the machine.

Thompson looked at his stopwatch. "Ready?" he said. "Cast mode." Those words notified the crew that the series of events leading toward a cast was about to begin. He looked at his stopwatch. "Slidegate open," he said, punching the stopwatch with his thumb.

Pelfrey, the ladle man, climbed a ladder and threw a lever below the ladle, and the slidegate opened. That was supposed to start the flow of steel into the bathtub.

Thompson pretended that the slidegate was clogged with a skull of steel.

"Burn it open!" said Thompson.

Dick Haltom, a steelworker from Crawfordsville, drove an oxygen lance up into the slidegate, fished the lance around, and then withdrew it and scurried away. Imaginary steel began to boom through the slidegate into the bathtub.

"Close the slidegate," said Thompson.

Pelfrey threw a lever and imaginary steel stopped running into the bathtub.

"Open the slidegate again," said Thompson.

Non-steel again began to roar silently into the bathtub. The bathtub filled rapidly with nothing.

"Temperature," said Thompson.

Haltom jumped up onto the bathtub and slid a probe into the bathtub to take the temperature of the bath.

"Open stopper," said Thompson.

■

The bathtub stopper was opened. Steel rushed into the Kolakowski mold, the mouth of the machine. The mold began to fill with imaginary steel.

Thompson hurried across the deck and knelt beside the mouth of the machine, eyeing his stopwatch. He counted out loud, "One, two, three . . ." until he had reached twelve. "Mold powder," he said.

At those words, Ken Kinsey, steelworker, had to push a load of secret gray powder into the mouth of the machine using a wooden hoe. The powder was supposed to lubricate the walls of the mold, to prevent molten steel from sticking to the copper. Kinsey scrabbled his hoe across the top of the machine, pushing piles of imaginary secret gray powder into the copper mouth of the machine.

Thompson called out, "Strand start."

At the words "strand start," Dan Harmon, steelworker, punched a button on a control panel, and a plug was pulled out of the bottom of the mold. The plug pulled a lengthening ribbon of imaginary steel out of the mold. The plug traveled straight down through the ladder of rolls, dragging the ribbon after it. The plug was disconnected from the ribbon of steel, and the ribbon was bent and withdrawn sideways from the base of the casting tower. The ribbon, or strand, could be pulled from the machine indefinitely, as long as liquid steel was being poured into the mold. Meanwhile the oscillator motor drove the mold up and down in a jogging motion to prevent liquid steel from sticking to the mold.

Thompson knelt on top of the machine while he gave orders to his men. The top of the machine was jumping up and down. Thompson's body jogged up and down. Genuine water began to *whoosh* through the machine. Thompson kept his eyes on his stopwatch. I recalled the words of Manfred Kolakowski in Germany: "All decisions have to be made in seconds."

The top of the machine humped faster—*chug-a-chug-a-chug-a* . . . *bonk, bonk, ka-bang!* The machine stopped.

The steelworkers looked around.

"What the hell was that?"

"We had an emergency shutdown of some kind."

Jim Hoskins, the scrapyard man, had dropped by to watch the practice. He suddenly shouted, "Breakout! Breakout!"

The steelworkers looked annoyed with Hoskins.

"You need another quarter," said Hoskins. He pulled a quarter from his pocket. "Here—here's a quarter. I cain't find the slot."

The machine had spontaneously shut itself down, causing a dry breakout—or rather, there would have been a breakout if there had been hot metal inside the machine. Nonexistent hot metal drenched the guts of the machine, and no smoke and no flames billowed across the deck, while genuinely nervous German engineers began to hurry back and forth across the deck. One of them peered into the mouth, muttering: *"Was ist los?"*— "What's happening?"

Kinsey sauntered over to me, carrying his powder hoe over his shoulder. "Don't attempt to cast steel at home," he said. "It can be done only by professionals. Or by rank amateurs."

■ ■ ■

The stopper in the bathtub, which governed the flow of steel into the machine, was moved up and down automatically by the casting machine's brain, a computer known as a logic controller, which was located in a room inside the casting tower. The logic controller kept going haywire, causing the bathtub stopper to do strange things like open too wide or clamp shut. If that happened during a real cast, too much metal would suddenly rush into the machine, or too little metal would go into the machine, and in either case there would be a breakout. They decided to set up an emergency manual override in case the bathtub stopper went crazy.

Here was the procedure. Thompson would say, "Go manual." Then a steelworker by the name of Steve Stout would grab a lever attached to the bathtub, and he'd use it to open and close the stopper, attempting to control the injection of steel manually. It had never been done before, not even by Manfred Kolakowski himself. In Germany, on the prototype machine, a computer had always controlled the injection of steel into the machine. Thompson practiced emergency manual overrides followed by an aborted cast, so that the crew could get a feeling for what it was like to have to take over manual control of the machine in the middle of an accident.

"Emergency procedures. Do it again," said Thompson. "Cast mode! Open slidegate! Temperature! Open stopper! Mold powder! One, two, three, four . . . strand start!"

The machine humped up and down, dry-casting steel. After a few moments Thompson shouted, "Go manual!" Stout grabbed the injection lever. They pretended to have a breakout. "Stop!" yelled Thompson. (Imaginary smoke and flames were now boiling up around the mold.)

Stout yanked up the lever to stop the flow of steel.

"That's good," said Thompson. "We're on a roll today." He was

■

nervous. The machine was within hours of being ready for steel. "This is like the first football game I ever played," he muttered to Millett.

■ ■ ■

That same day, Thursday, June 15, 1989, at five o'clock in the evening, Keith Busse paces his office. He believes that the casting crew will attempt to cast tomorrow. He decides to try on his silvers—splash-resistant aluminized clothing worn by steelworkers on a casting deck. He wraps a silver bootie around his calf and fastens it with Velcro tabs.

"This is supposed to cover my boot?" he says, tugging a flap of aluminized fabric over his boot. "Is that the idea?"

He puts a Styro cup of black coffee to his mouth, swigs it, and walks around his office with a silver bootie wrapped around one foot, and then he rips the bootie off his leg and tosses it into a pile of silver clothing on a chair. He unlaces his steel-toed boots and puts on his tasseled loafers.

Busse isn't going to stick around the plant tonight. "I'm going to watch my son Aaron play baseball," he says. "Tonight I'm going to be Dad for a little while. Aaron's pitching two innings in the Little League tonight. I don't even know who they're playing. I haven't had time to find out."

■ ■ ■

It is late on Thursday night, the night before the first cast. The casting deck swarms with people. A crane hovers over the mouth of the casting machine, and from the crane hangs a long ribbon of steel, an artificial slab. They lower the slab through the mouth of the casting machine. The slab travels down through the machine with cracking noises and strange groans.

"It's bumping the whole way," mutters Thompson.

Millett kneels and shines a flashlight down into the machine. "That looked pretty good to me," he says. Mechanically the casting machine seems to work. At least it works with a piece of solid steel.

"This is like the night before Christmas," says Millett to Thompson. "The night before the first cast. If we cast tomorrow, are you going to wield a bottle afterward?"

Big Dave Thompson offers a thin smile. "I have a feeling my head will be crushed."

Thompson isn't confident about the machine. He wants an extra day to practice dry casting. He takes his fellow foremen into a huddle and says, "If you want to know the truth, I'd feel more comfortable if we spent Friday drilling the hell out of ourselves and then casting on Saturday."

■

207

He sits down beside the copper mouth of the machine. He takes off his hard hat and rubs his hand down the back of his head, pulling sweat off his short, damp hair. "We've wanted to get a slab through this machine in the worst way," Thompson explains to me. "But I'm crawly by nature in situations like this. Most of us have downplayed the fact that this is the first one of these machines. But then we think, 'Gee, the whole world is watching over our shoulder.' But our job is to set up the caster and make some money for ourselves. We want to better ourselves. That's why we are here."

Millett gathers the crew for a brief talk. "I think it will work tomorrow," he says to the crew. "And if it doesn't work, it won't be for lack of trying. It all depends on the stopper control"—referring to the injection system. "That's the one thing you can't simulate. Molten steel is the only way to test it."

The casting crew nods. They are chewing peppermint gum. The air reeks of peppermint.

"If we have a problem with it," says Millett, "we'll go immediately to manual control. Then we'll punch new buttons. We'll try new settings. We'll try every button we can think of. If nothing works, we'll terminate the cast." He gives them a thumbs-up sign. "O.K., mates. Go on home and get some sleep."

"I won't sleep tonight anyway," says Ken Kinsey, steelworker.

As the crowd breaks up, Millett turns to Thompson and says in a low voice, "It's your show tomorrow."

Millett will try to get some sleep, too. "Abby hasn't seen much of me lately," he remarks. He descends the exit stairs from the casting tower and walks slowly through the repair shop, among duplicates of the casting machine, among gleaming pieces of new industrial equipment, untouched by fire. Millett passes through the east door of the melt shop into the night air. It is near midnight, a cool night without wind. Here and there in Montgomery County, employees of the Nucor Corporation are trying to get some sleep. F. Kenneth Iverson, in his house on the outskirts of the city of Charlotte, North Carolina, knows that the first cast is coming, and he also may be having trouble getting to sleep, because pressures at the company tend to keep Iverson awake at night. Iverson has no intention of visiting Crawfordsville during the startup. He stays away from startups for the same reason that he keeps the corporate headquarters in the Cotswold Building, at a safe distance from any Nucor steel mills. As he once expressed it, "You don't put the corporate headquarters near a plant. That's like having your mother-in-law next door." He wants to let Busse, Millett, and the others rattle around in their own lonely cages.

A waxing moon stands over the melt shop. Along the eastern horizon a few lights burn, farms along the headwaters of Cornstalk Creek.

"For the sake of everyone's sanity, it had better work tomorrow," remarks Millett. He climbs into his rust-colored Pontiac Grand Prix and the door slams, an echo of flat-rolled steel.

15

A HEAT
IN THE
MACHINE

■ ▪ ■

The day scheduled for the first cast, Friday, June 16, broke cold. At seven o'clock in the morning, the casting crew gathered before the door of the melt shop, where Big Dave Thompson, the helmsman and quarterback, spoke to the crew. "It'll be a while before we get a heat of steel," he said. "We want to go through more simulations. Do 'em as fast as you can. These are the pregame festivities. Show 'em the play we want to run. Full dress rehearsal. In silvers."

"Silvers? This ain't no fashion statement," said a steelworker.

"You keep your mouth shut, boy," said Thompson, and the crew grinned.

A foreman by the name of Dave "Smitty" Smith was a little worried about Franz Küper. "We've got to keep Franz out of the pulpit," said Smitty. "He slid in there yesterday and punched a button. Dan Harmon said to him, 'What the *fuck* are you doing?' "

Guffaws from the crew.

Smitty grinned. "You never know what these Germans are going to do next," he said.

Someone's walkie-talkie crackled and said, "T-minus five hours and counting."

"Let's go!" said Thompson.

The men crowded into the melt shop in a herd, ran up the stairs of the casting tower, and stood around nervously on the deck.

"Where's the furnace? I don't hear it yet."

■

"It ain't started yet."

"Long as I can get drunk tonight."

Miller, a casting foreman, said to Thompson, "You think you're going to want a cocktail after this is over, Dave?"

"I'm going to bed," said Thompson.

"Hell, if you lie a lot, Dave," said Miller.

"It took us more than seventeen seconds to charge down those stairs," remarked a steelworker.

"*You* couldn't get down those stairs in seventeen seconds unless you rolled," said another.

"Yeah, and we got a wheelchair ramp for you."

A rumble infiltrated the melt shop: the south electric arc furnace struck an arc. The men cheered, nervously.

Franz Küper paced the casting deck, wearing a white hard hat. His gold-rimmed spectacles glinted and his mouth was tight. "Anything can happen today," he said. "I am not surprised by anything anymore."

Patrick Owens, the commissioning engineer, sat at a video screen in the pulpit. He patted his hands across a pack of Merits, saying, "I got my cigarettes, I got my furnace glasses, I got my silver suit, and if this machine goes into a steady state, I'll put on my silvers and watch it make steel. And I'll sober up on Monday."

The furnace died, and a hush fell over the melt shop.

Someone said, "The breakers are tripping out again."

Manfred Kolakowski walked the deck from one side to the other, stopped by the mouth, knelt down, and reached inside the Kolakowski mold. Mark Millett pointed a flashlight into the mold so that Kolakowski could see what he was doing.

Suddenly there was a roaring hiss. A white jet of fluid erupted from a wall near the west end of the casting deck. It had nothing to do with Kolakowski, and so Kolakowski paid no attention to it.

"What happened? What happened?" said the steelworkers, milling around.

Mark Millett hurried to the scene of the accident, and returned to report that a hydraulic valve had blown a pinhole leak. A stream of hydraulic fluid spewed from the pinhole at 2,700 pounds per square inch, in a thread as thick as a pencil lead. It jetted forty feet through the air and at the end of its fall it looked and sounded like water coming out of a firehose. It could have sliced through skin like a kitchen knife. A foamy slick of hydraulic fluid coated the casting deck.

■

"Hose the fluid down with water," said Millett, giving orders to the men. "That's flammable stuff. Blow it off the deck. Do it quickly. We want to keep the water on the deck to a minimum."

A mixture of water and hydraulic fluid could soak into the concrete deck, and then if any liquid steel dropped on the deck, the deck could explode.

The furnace resumed operation with a drumroll. The furnace men were melting at near-full power today, at sixty-five megawatts.

No sooner had Mark Millett got the hydraulic leak under control when the lights went out—the power failed throughout the entire Crawfordsville Project. Sixty-five million watts went off line. Silence fell over the melt shop as the electric arc furnace stopped melting steel.

"I don't know what's happened," said Millett, looking around. Suddenly his walkie-talkie began to shout: "Mark Millett! Mark Millett!"

Emergency on the furnace deck.

Millett took off at a dead run for the arc furnace. He ran downstairs, through the nave of the melt shop, and up a flight of stairs to the furnace deck, taking the steps three at a time.

■ ■ ■

Water had stopped circulating inside the walls of the arc furnace when the power died, a dangerous situation, because the furnace could overheat and explode. Fearing that the furnace was about to pop, the furnace men had abandoned the furnace deck. The furnace hadn't popped, and now they were edging their way back to the deck along a walkway to study the situation.

"That son of a bitch is hot," muttered a furnace man, squinting at the furnace.

"We're not off to a good start today," said another.

The roof of the furnace had been raised about six inches, but because of the power failure, the roof was now locked over the furnace vessel, and could not be moved to one side to release the accumulated heat inside the furnace. Molten steel illuminated the underside of the roof. James Mc-Caskill, harpooner, climbed a ladder and peered over the lip into the furnace. Millett went into the pulpit to discuss the emergency with the Swiss furnace experts.

"I'm getting a little antsy for that roof to move away," said a furnace man.

■

"You can get a vapor lock in that son of a bitch. Then when you start it up, the furnace'll blow a panel."

A furnace man took me aside and told me what to look for if the furnace was getting ready to explode. "Keep your eye on James McCaskill," he said. "If you see James running, then you just follow him. And if you see any *Germans* running, get out of the building and don't look back."

Abruptly the roof of the furnace edged away from the furnace vessel, under auxiliary power. A puff of accumulated heat boiled out of the pot. Millett hurried out of the pulpit, and just then, the lights came back on. With a *whoosh,* cooling water began to circulate inside the furnace. "Excellent!" said Millett. "It didn't blow up!"

Millett ran back to the casting tower, where he found Dave Thompson and said to him, "It's going slowly at the furnace. It'll be a long day. It'll be four or five o'clock this afternoon before you get a heat of steel. Get the key players together and have a chat with them about all the procedures. You'll need to eat, too. The way things are going, we may not cast until midnight."

Owens, the commissioning engineer, remarked to Millett, "I have not been sleeping. I've been passing out for a few hours, and then I get up."

"I was a log last night," replied Millett.

"All the emergency functions have been tested," said Owens.

"Good, because we'll need them all," said Millett, grinning.

■　■　■

Morning gave way to afternoon, and Keith Busse went to visit the cold mill building. He found a crowd of cold mill workers standing around with their hands in their pockets at the base of one of the used Austrian rolling machines, a Vöest-Alpine reversing mill. A reversing mill can take a strip of hot band steel the thickness of rawhide and three thousand feet long, and can thin it and elongate it into a sheet that is miles long and a few hundredths of an inch thick. A reversing mill does this by rolling the steel back and forth through a single set of rolls, like film being run back and forth through a camera. The reversing mill was a single millstand that rose three stories from the floor and extended one story below ground. At the moment nothing was happening, except that water was gushing down through the machine and into the basement of the cold mill.

"It makes a nice waterfall, anyway," said a Nucor man, with his hands in his pockets.

■

The cold mill people had attempted to roll some coils of steel in the reversing mill, and the results sat on the floor—three wrinkled coils of sheet steel with ragged edges. The coils looked like rolls of toilet paper that have been unraveled and rolled up again. Busse had purchased the coils from the U.S. Steel Gary Works, on the Lake Shore. Having been passed through a Nucor reversing mill, the metal was now Distressed Material.

Keith Busse huddled with a maintenance man and the cold mill manager, Vincent Schiavoni.

"The only constant with this cold mill," remarked the maintenance man, "is Westinghouse. You can be sure if it's Westinghouse. You can be sure it won't be ready on time."

"I didn't like that power failure this morning," said Busse, changing the subject. "We even lost our telephones. I want you to fix the emergency generator so we don't lose our telephones."

"We could buy a nuke," said the maintenance man.

"Nice idea," said Busse.

"Seabrook is for sale, in New Hampshire," remarked Schiavoni, the cold mill manager.

"Probably get it cheap," said Busse, and then, "How much longer before they tap the furnace? Have you heard?"

"Thirty minutes, about," said the maintenance man.

"Can I borrow your walkie-talkie for a minute?"

"Sure."

"Mark Millett! Mark Millett!" said Busse.

"Yes," came Millett's voice over the walkie-talkie.

"What do you mean you haven't tapped the heat yet?"

"We're waiting on you!" crackled Millett.

Busse grinned. "I'll be there," he said.

Busse walked upstream along the manufacturing line, heading for the pool of hot metal. He walked the length of the cold mill building, past the antique Austrian cold-rolling machines. He crossed an open space and went into the hot mill building, where the four millstands of the CSP stood motionless under floodlights, their guts exposed to workmen, who clambered over the millstands on catwalks. Welding flares gleamed in the machinery. The lower half of the CSP was a mess. It was not ready for steel. Trouble in the cold mill, trouble in the hot mill, trouble everywhere.

He headed for the furnaces, where he found Mark Millett.

"So what's the news, chief?" he said to Millett.

The news was bad. Millett confessed that he had accidentally over-

■

cooked the steel. His people had blown too much oxygen into the bath while lancing down the heat. That had burned all the carbon out of the metal. Millett had made an ultralow carbon steel—almost pure iron—unintentionally.

Busse was irritated. If you poured pure iron into the caster it would break out for sure. The machine might puke. Millett had burned the soup, and it did not amuse Keith Busse. Possibly Millett was going to have to dump this soup onto the floor and start over again with fresh ingredients.

Millett conferred with the furnace men to see if he could resurrect his soup, while Busse paced the deck with his hands in his hip pockets.

"They've been pattycakin' around and now they've screwed up!" Busse muttered to me, on one pass. On the next pass he said, "They done played with the steel too long and they burned all the carbon out of it!" And on the next pass: "They tell me it's awfully hard to put carbon back into steel once they've taken it out!" Next pass: "They should have tapped the heat when they had the carbon right, instead of pattycakin' around! They should have tapped the heat and sent the steel downstream and *forced* the issue at the caster!" The arc furnace crackled. "Ultralow carbon!" he shouted to me, over the roar. "They pay big money for steel like that! Only problem is, we're not trying to make that kind of steel today! And Mark knows that!"

Millett came up with a proposal to dip the graphite electrodes into the liquid steel. He thought that might resurrect his soup. Dipping the electrodes into liquid steel would burn up the electrodes, but they were made of graphite, pure carbon. The electrodes would dissolve in the bath, adding carbon to the bath, turning it into a mild steel.

They kept the furnace door open to watch the dipping operation. The electrodes descended and touched liquid metal. There was a chatter of arcs, sparks zoomed out through the slag door, and the electrodes dipped a few inches into the bath. The noise diminished to a deep hum. Electrodes buried in steel don't make noise. They short out on the power system. A minute later the electrodes were withdrawn from the metal with a horrifying crackle.

James McCaskill, harpooner, examined the electrode tips through dark glasses. "We just ate up fourteen hundred dollars' worth of electrode," he said.

A sample revealed that it was a mild carbon steel; Millett had rescued the soup. At 4:30 in the afternoon, Millett declared a finished heat of steel. The furnace men tilted the furnace, opened the tap hole, and a bright,

■

heavy, liquescent pour of mild steel rumbled into a ladle below the furnace. The crane carried the ladle up to the metallurgy deck to be kept warm and to wait for Big Dave Thompson to call for hot metal.

■ ■ ■

On the casting deck among the crowds, a salesman from Steel Grip, Inc., out of Danville, Illinois, displayed cardboard boxes containing brand-new silvers, fireproof space suits for the casting crew. The casting crew helped themselves to silver overcoats, silver neck shields, silver gloves, silver booties, and clear faceshields. When they donned the gear they looked like astronauts. They walked around grinning, with their hands stuck out from their sides.

The silvers had an aluminized outer layer and a core of Kevlar and carbon-graphite fibers. In theory the silvered material would shed molten steel the way that oiled paper sheds water.

"We puddled some steel on one of these," said a steelworker. "It didn't burn through."

"I'll skip the experiment," said Kinsey, steelworker.

The casting crew sat in a row on a concrete ledge in their silvers, drinking cans of Mountain Dew.

Keith Busse appeared, wearing silvers. "You guys are all suited up like spacemen," Busse said to them. "You going to do something important today?"

"We kind of hope so," said a steelworker, tipping his Mountain Dew.

"She'll either make steel or she won't make steel," said another.

A steelworker said to Busse, "Are you getting nervous?"

"Nope. I have confidence in you. And in the machine. Hey, Kinsey!" he said. "There's a rumor going around they're going to ship your ass back to Vulcraft in Saint Joe, where you'll be a rigger on a rigging table."

The men in silvers laughed.

"I hadn't heard that one," said Kinsey.

A steelworker said to Busse, "What are you going to do with the slab once you get it out of the machine? You just going to drive it through the millstands?"

Busse grinned. "I think Rodney Mott would be upset." He was referring to the fact that Mott hadn't finished work on the millstands of the CSP, in the hot mill, and that if a slab were pushed into the millstands now, there would be a horrible wreck.

■

Busse added, "If we get a slab, we'll run it down to the end of the tunnel furnace and cut it up with torches." He walked off with Millett to talk about something.

The steelworkers seemed a little less eager when Keith Busse was out of sight. "I'm so tired, I could probably drink two beers tonight and then pass out," remarked one guy.

"I think we're gonna cast in about an hour," said another.

"One hour from now we're either gonna be heroes or turkeys."

"Yeah, it's gonna be either 'Attaboy!' or 'Uh-oh!' "

Said a steelworker to Ken Kinsey, "I'll give you ten bucks if you say to Busse, 'Hey, Keith, come over here and push some powder.' Ten bucks. Ten bucks if you tell Busse he can push powder into the mold."

"I already told him that," said Kinsey. "I said, 'I'm going to step back and let you take my place.' "

"What'd he say?"

"He said, 'Don't tempt me.' "

■ ■ ■

At seven o'clock in the evening, a squall line appeared to the north of Crawfordsville, stormboil moving along the Lake Shore, and the Rust Belt turned away from the sun into space. It was a sweet June evening, near the summer solstice. Flycatchers dodged around the melt shop. The moon budded in the east, as red as a berry, then as white as hot metal. On the New York Stock Exchange, Nucor common had closed the day at 56¼, up ⅞ of a point on heavy volume. No one on the casting deck ate dinner; no one wanted to eat solid food. The crew drank Mountain Dew and smoked cigarettes for their dinner. Some of the men had worked at the plant all night the night before; they had now been awake for thirty-six hours.

I put on a set of silvers. First I wrapped the booties around my legs—they resembled Frankenstein boots, with a spat that covered the foot—except that my toes were exposed. I decided that I could spare a few toes if a splatter hit my feet. The silver coat was too short and did not cover the groin sufficiently for my taste. I tried another silver coat. It was much too large for me, but at least it hung below the knees. The silvers were surprisingly light and airy, and they crinkled when I moved. It was difficult to believe that even though the material was woven from Kevlar and graphite it could stop a splash.

For your neckshield you could choose between leather or silver. Most of the steelworkers chose leather. The neckshield was supposed to hang from

■

the base of your hard hat like a sort of wig, to deflect any droplets of steel that might fall from above and keep them from running down your back. I couldn't figure out how to attach the neckshield to my hard hat. A steelworker showed me how to fasten it. It draped over my shoulders. "Now you look like you're in the Foreign Legion," he said.

To the brim of my hard hat I clipped a set of smoked furnace glasses, for observing the steel.

Hans-Friedrich Marten, the SMS executive who had negotiated the deal for the CSP with Iverson, appeared on the casting deck for a quick inspection tour. Franz Küper accompanied Marten. Marten peered into my face, under my hard hat and neckflap and smoked glasses. "What are you doing here?" he said sharply. "Who said you could be up here? This is very dangerous. Steel can blow up."

A siren went off. The siren distracted the German executives and I edged away from them. They left the casting deck. It was 7:25 P.M.

Manfred Kolakowski smiled at the sound of the siren. He crossed the deck to a locker and put on a set of silvers. Mark Millett, also wearing silvers, shook hands with Kolakowski. "Good luck, Manfred," he said.

Kolakowski said, "When you are with child, we say in German you are 'in good hope.' We are in good hope tonight." He went off by himself to a corner of the deck and paced alone, with his hands behind his back, staring at the floor, staring into space, in good hope.

The siren grew louder. It signaled the movement of the ladle full of steel toward the casting deck, carried by a traveling crane that was moving along the roof of the melt shop. A red warning light, like a police light, began to flash above the deck, illuminating the men's silver coats with red reflections, on-off, on-off, blood-silver, blood-silver.

At 7:55 P.M., the steelworkers seemed to pull into themselves.

"You got a cigarette?"

"We got time for one more? Yeah?"

"Heat's gonna be here in five minutes."

On the floor at the base of the casting tower, a crowd of Nucor employees gathered. A cluster of faces appeared on the exit stairs—people not wearing silvers, but who were anxious to watch the casting crew pour metal into the machine. Four hand-operated fire extinguishers were lined up on the deck. It didn't look like they would do any good if a big steel fire got going on the casting deck.

Kinsey rolled a wheelbarrow over to the machine. The wheelbarrow held a plastic bag full of secret gray mold powder, to be fed into the machine along with molten steel, in order to lubricate the metal as it moved through the machine. Kinsey slit the bag and dumped its contents in a heap beside the mouth of the Kolakowski mold. "I've got mold-powder fever," he muttered.

Stout, the manual injection man, wrapped his silvered gloves around the injection lever, which was attached to the bathtub, and he moved the lever up and down, feeling the stopper move inside the drain hole in the empty bathtub.

Thompson pulled a telephone receiver off a pier and spoke to Owens, who sat inside the pulpit at his colored video screen, poised to operate the machine logic. Then Thompson turned and said something to Kinsey.

Kinsey nodded and turned to Stout: "Steve," he said. "We're going to do a test. We're going to open and close the stopper, up and down, once."

This was a test to see if the logic controller could control the bathtub stopper.

Stout removed his hands from the lever.

The lever waved up and down. The casting machine was under automatic logical control. Owens, in the window of the pulpit, gave an O.K. sign with finger and thumb.

Keith Busse, in his new silvers, stood alone at the west rail of the casting deck, watching the crane bring the ladle of steel the length of the melt shop toward the casting deck. The ladle contained one hundred tons of molten steel. A glow came from the top of the ladle, flooding the roof of the melt shop with a fuzzy disk of orange light. Busse put his hands on the rail. No one went near Busse.

A gas burner was roaring inside the bathtub, heating it to receive steel. Steelworkers turned off the burner and removed it from the bathtub. The interior of the bathtub glowed a dull red.

The crane deposited the ladle on a ladle car at the west end of the casting deck. The ladle car began to inch toward the mouth of the machine, more sirens went off, and the police light turned around and around. The deck shuddered slightly under the weight of a hundred tons of steel.

A row of entranced faces inside the pulpit tracked the ladle as it moved slowly toward the machine. A little black pipe stood out from the bottom of the ladle: the sapphire slidegate, the muzzle of the gun. The ladle halted over the bathtub.

Off to one side, Millett positioned himself next to Kolakowski. Millett wore his red checked flannel shirt.

Thompson turned around and caught Millett's eye. Millett nodded. In

■

that silent glance Millett handed control of the casting deck away to Thompson. From now on, Thompson would make all the decisions. Thompson had taken the helm of the machine.

Thompson picked up the telephone off the pier again, to speak with Owens in the pulpit. "Do I have everything I need?" he said to Owens.

Owens's eyes moved over colored symbols on the screen as he went silently through the final machine checks: *Machine logic, O.K. Mold water, O.K. Machine water, O.K. Spray water, O.K. Main hydraulics, O.K. Stopper hydraulics, O.K.* Owens said into the telephone, "You have everything you need, Dave."

Thompson crossed the deck and got down on his hands and knees at the mouth of the Kolakowski mold, where the white nozzle poked between the copper lips. He inspected the nozzle for cracks. A broken nozzle could clog the throat of the machine, causing the machine to vomit steel across the deck—the same kind of accident that had troubled the pilot plant in Germany.

Thompson stood up and beckoned to his men. They gathered around him in a huddle. "Everybody ready?" he said, looking around.

Nods here and there.

"Whoa!" Thompson yelled.

The steelworkers roared at the top of their lungs, broke the huddle, and took up their stations. They lowered their faceshields.

Busse crossed the deck and placed himself behind Kinsey, six feet from the mouth of the machine, where he could watch the cast at close range. Busse lowered his faceshield. Kinsey crouched beside the nozzle, holding his hoe ready to push powder into the mouth on Thompson's order.

Thompson said: "Position shroud."

Pelfrey, the slidegate man who had formerly worked in the book factory, fitted a white sleeve over the small black muzzle of the ladle slidegate, at the base of the ladle.

A moment later, at 8:10 P.M., Thompson said in a calm voice, "Cast mode." The sequencing had begun, but nobody moved. All eyes were on Thompson.

Thompson turned 360 degrees on his feet, to make sure that his men were in position. "Everybody ready?" he said.

The crew nodded.

Thompson picked up his telephone receiver. "Everybody ready in there?" he said to the pulpit.

"Yes."

"Open slidegate!" he shouted.

The black pipe at the bottom of the ladle moved sideways, two holes lined up, and steel came through. It was a full-stroke free opening. A hiss and a warm glow shot from the muzzle. The hiss mounted to a roar. A wave of light and heat washed over us, cooking our faces. The metal emerged from the slidegate at breathtaking volume, a rocket nozzle firing into the bathtub. The bathtub filled rapidly, and as it filled, the metal rolled in the bathtub and howled and seemed to be trying to escape. The bathtub popped and crackled and groaned under the rapidly increasing weight of the steel. A golden light spilled out of the top of the bathtub and illuminated the bottom of the ladle.

Two steelworkers jumped up on the bathtub car, carrying bags of charred rice hulls. They tossed the bags onto the steel in the bathtub. The bags split open and the rice hulls formed a tight black powder over the metal, protecting it from the air. Flames leaped from the top of the bathtub and licked at the steelworkers' sleeves.

"Temperature," said Thompson.

The temperature man, Dick Haltom, stabbed a lance deep into the bathtub to take the temperature of the metal. The lance failed to sense any temperature. Haltom yanked out the lance and threw it across the deck, impatiently, and it bounced away in a rain of sparks. He grabbed a new lance, put it into the bath, still got no temperature, and threw that lance away. Steel continued to rumble through the slidegate into the bathtub, one ton every eighteen seconds. Haltom couldn't get a temperature reading, and the bathtub was filling up quickly, threatening to overflow and flood the deck.

Thompson considered aborting the cast. He didn't want to abort. He decided to press on with the cast, but to try to throttle back the flow of hot metal into the bathtub before it overflowed, to allow Haltom time to get a temperature reading. "Cut it back!" he said.

The slidegate man worked a lever. The full-stroke pour was reduced to a drool.

Miller, a foreman, studied a gauge. "Twenty thousand," he called out. Those words meant that twenty thousand pounds of liquid steel sat in the bathtub.

The bathtub was a little bit overfilled. They had been aiming to fill it with eighteen thousand pounds of hot metal—nine tons—before they started the cast.

Haltom shoved a third probe into the bathtub, and finally, on an illuminated board above the pulpit, the temperature of the steel appeared in red digits: 2,830 degrees Fahrenheit.

Thompson turned on his feet and studied the number. The steel was a little bit cold. Molten steel that was too cold might jam inside the machine, triggering a breakout. He had only two alternatives: either to abort the cast or to put hot metal into the machine and take his chances. He had to make up his mind fast, because the steel was cooling off even as it sat in the bathtub. He decided to put metal into the machine. First thing was to get the slidegate open again.

"Open slidegate," he said.

A hiss came from the ladle, and the flow of molten steel resumed into the bathtub. The light struck our faces again.

Big Dave Thompson wanted steel inside the machine *now*. He grabbed the telephone and said, "Stopper open."

Owens, in the pulpit, pressed a button, ordering the logic to open the bathtub stopper. Nothing happened.

"We're on auto," said Owens to Thompson, over the telephone. "It's trying to open and it's not opening."

Thompson's mind was racing. He realized that the stopper was welded to the bathtub's drain hole by a lump of steel, a skull. He realized that the delay in taking the temperature had allowed steel clots to form around the drain hole. The stopper was frozen shut.

Meanwhile the bathtub continued to fill up with steel. The bathtub was now groaning with hot metal, one ton, eighteen seconds, one ton, eighteen seconds. A familiar skunky smell of hot slag and liquid steel filled the air.

"Twenty-five thousand," cried Miller. Twenty-five thousand pounds of hot metal sat in the bathtub. It was threatening to overflow.

To hell with the computer. *"Manual!"* shouted Thompson. *"Go manual!"*

Stout, the injection lever man, threw his weight on the lever. Nothing happened. The stopper was welded tightly into the drain hole.

Thompson gave a hand signal, and two steelworkers joined Stout, the three of them putting their weight on the lever, wrestling with it. The stopper came open with a jerk.

A bright flash occurred at the nozzle, inside the copper lips. There came a cough and a hiss, and the nozzle disgorged steel into the machine. A yellow-white glory of clear beautiful light streamed out of the mold and around the men in silvers, who knelt in the light. They appeared to be praying. They were only doing their jobs. And then there came the sound.

■

Because the mold had a curved copper surface, it focused the sound like a horn. The horn gave off a deep sucking bubbling metallic *whoosh* as it filled up with steel.

Stout, the lever man, discovered something new. He discovered that he could not control the lever. It dragged and bucked against his grip, tugged by the sheer weight of the steel running through the drain hole into the mold.

Thompson shouted the countdown over the roar: "One! Two! Three! Four!" At twelve he shouted, "Powder!"

Kinsey began to push powder into the mold with his hoe. Busse crouched behind Kinsey, peering into the mold. Metal squirmed brightly in the mold.

"Fourteen! Fifteen—!" cried Thompson.

The mold now contained nearly one ton of hot metal.

A warning light on a panel near Dave Thompson flashed on. It read, "Mold Level Problem."

Stout had lost control of the lever. He jerked it up and down.

The emergency panel lit up with lights.

There was a hiss and a wild eruption of sparks. The mold caught fire and boiled. The machine vomited up through its mouth, and Kinsey and Busse, who were leaning over the machine, fell backward. The mouth coughed a burst of droplets.

The steelworkers froze in terror, their forms silhouetted in a brightening glare coming from the machine's mouth, their arms outstretched or bent, covering their faces, bedazzled by sparks.

"*Shut it off! Shut it off!*" Thompson shouted.

They couldn't shut it off.

Accidents with molten steel are instantaneous. In the blink of an eye, we were enveloped in sparks. The sparks soared out of the machine. A gout of flame burst neck-high from the mold. Bits of hot steel were flung from the machine. Steel raced out of the mouth and bubbled onto the casting deck around the steelworkers' feet. The steelworkers danced backward. They paused, found their nerves, and went on the attack. They rushed into the flames, hopping and leaping around rivulets of steel.

The nozzle was still injecting metal into the machine, because Stout, who had clung to the lever, could not get the stopper closed to shut off the pour. Two men joined Stout. Together they pushed the lever up as far as it would go and held it there, and the rushing of steel into the mold abated. Pelfrey, on the far side of the deck, stroked the ladle slidegate shut. Fortunately the slidegate closed, clipping off the pour into the bathtub. Long whoops of a siren sounded overhead. The crane snatched the ladle

■

into the air and away from the casting deck. The cast had been aborted.

The mold continued to erupt and burn, even after all movement of steel through the system had been halted. Then the steel fire in the mold died down. A cloud of smoke hung over the casting tower. It was a successfully aborted cast.

The men flung off their silvers, in profound discouragement.

"What happened?" I asked a steelworker.

"Just a mess to clean up," he said.

Parts of the casting deck were slobbered with twisted, blackened ropes and puddles of skull. Steel fires muttered on the deck. A smell of cooked paint filled the air, and the skull let off an odor of cat piss. The Kolakowski mold was clogged with a huge, heaped skull—a lumpy red stalagmite of metal, breathing flames. The nozzle was buried in the metal, locked into the mold. The men smashed the nozzle with a sledgehammer to detach the bathtub from the mold.

The inventor had gone off by himself to the end of the deck, where he paced alone.

Mark Millett climbed onto the bathtub car and stared down into the hot metal, his face glowing with blackbody light coming from the bath. He seemed to be staring into the center of the earth. He thrust a steel pipe into the bath and idly stirred it, considering what to do next, while the pipe melted away.

Then a water leak developed in the mold and clouds of steam boiled up from the mold, dissolving Millett's outline. Gray mold powder was strewn all over the casting deck, mixed with bits of steel. It crunched underfoot. The casting crew took up wrecking bars and viciously chopped steel away from the casting deck and the mold. They were visibly angry.

"Can you save the mold?" I asked one of them.

"Sure we can. Every mold is made for breakouts."

Owens, in the pulpit, pressed buttons and entered keyboard commands. Soon the mold's bottom plug moved upward. As the plug came up, it pushed the skull out of the mouth of the mold. The steelworkers cut notches in the skull with cutting torches. They fitted chains to the notches, hooked the chains to a crane hook, and the crane dragged the skull free from the mold. The skull came up smoothly and dangled in the air. It was a lump of steel weighing one ton. It was a cast of the Kolakowski mold. It had curved, lens-shaped faces. It also resembled a bulging letter envelope. But it was a warty thing with a red, mean heart.

Kolakowski had come over to watch the operation.

"Our first slab," an engineer joked to the inventor. "The surface is a bit rough."

"I am happy now," Kolakowski said in a dry voice.

The skull was too hot to be placed directly on the deck. Some steelworkers found a couple of wooden beams and laid them side by side on the deck, and the crane laid the skull across the wooden beams. The wood immediately burst into flames—the temperature of the skull was about a thousand degrees.

Mark Millett had climbed down from the bathtub, and now he and Keith Busse huddled in conversation.

"What's the story, chief?" said Busse. "Pretty nice cast until it aborted."

Millett had had enough. He wanted to cancel the attempt for two days, at least until Monday, to give his men some rest and a chance to practice more dry casts, casting air rather than steel.

Busse wanted Millett to try for a second cast immediately. There was plenty of hot metal left in that ladle, Busse pointed out.

Millett resisted, upset with the failure.

Busse wanted Millett to put more hot metal into the machine, tonight. But Busse wasn't willing to give a direct order to Millett to proceed, because Millett was the manager of the melt shop. He told Millett that the decision was Millett's. But then he continued to push Millett. Three times he urged Millett to "go in and do it again." He said, "Once the decision is made to go out and attack the son of a bitch, you have to see it through to the end."

Dave Thompson joined the discussion. He said to Millett that his men did not want to give up now. They had come too far tonight; they couldn't turn back at this point. "Let's give it another shot, Mark," said Thompson. "The guys have lost a round in a fight and they're ready to come back into the arena for round two."

Millett nodded. "Let's go. Bring in a new tundish," he said, meaning a new bathtub.

■ ■ ■

A tundish is lined with ceramic boards to keep it from melting. If you put hot steel into a cold bathtub, the ceramic boards can "pop," throwing small droplets of steel through the air. So the bathtub must be preheated before steel can be poured into it. On the casting deck stood two steel structures that held twin natural-gas burners. The steelworkers rolled the

■

new bathtub under the gas burners, lit the gas, and left the bathtub there to cook for a few hours.

Someone called Domino's Pizza and ordered twenty-five pizzas for delivery to the Crawfordsville Project. In a conference room near Mark Millett's office, a dozen steelworkers and engineers waited for the pizzas to show up.

"Those pizzas should be here by now. All we got to eat in here is a bucket of fried chicken."

"I'll take a damn wing."

"Where did the Germans go?"

"They packed into a car and went into town to eat."

"They could be cursin' us out and we'd never know it."

"They were. Believe me, they were."

A Buick Cutlass showed up at the melt shop. Inside the Cutlass there were a man and a woman and twenty-five Domino's pizzas. The pizzas hit the casting deck at 11 P.M. sharp, and four minutes later they were gone. Some steelworkers ate slices stacked three on top of one another. The crane operator threw down a rope, and the steelworkers tied a box of sausage pizza to the rope. "You want Sprite or Pepsi?" they called up to him.

■ ■ ■

"I was a little nervous," Manfred Kolakowski would recall later. It was about midnight now, and he paced the casting deck with his hands in his pockets. The night had grown cold, and it was cold inside the steel mill, but Kolakowski wore a short-sleeved candy-striped shirt. He seemed oblivious to the chill. He stopped Millett and said to him, "As soon as the steel starts to flow, you have three seconds to adjust the stopper rod, or the mold can overflow." Kolakowski wasn't surprised that the mold had overflowed. The pilot machine in Germany had always been run by a computer. No one had ever been able to feed the machine by hand.

A little after midnight, the new bathtub had been heated and was ready for steel. The crane picked up the bathtub and gradually lowered it toward the mouth of the mold, the white nozzle protruding from the belly of the bathtub. Dave Thompson put his hands on the bathtub to guide the nozzle down into the machine. At 12:25 A.M., the nozzle entered the lips of the mold.

"Inch it down! Inch it down!" shouted Thompson, wrestling with the hot bathtub. It weighed seven tons. Suddenly the crane jerked. The nozzle

■

struck the mold and shattered like an eggshell, dropping pieces of white ceramic down into the machine.

There was a moment of stunned silence, and then a groan went up from the crowd.

"We're *done!*" said a foreman, spitting on the deck.

"We busted the nozzle!"

"I can't believe this!"

"Son of a bitch!"

"That's it," said Millett. "We'll try again on Monday."

Keith Busse hurried into the crowd of steelworkers. He said, "Is there any other tundish ready to go?"

No, there were no other bathtubs ready to go. The steelworkers told Busse that if they were going to cast tonight, they would have to prepare a new bathtub, first lining it with boards and then heating it in the gas flames. "We're talking four o'clock in the morning before we can get a new tundish," said a foreman to Busse.

Busse, Millett, and Thompson huddled. Millett was shaking his head. Millett had had enough. Some of the casting crew joined the group. The talk became impassioned. The casting crew was arguing with Millett. Busse joined the argument. The casting crew had refused to go home. Millett could order them to go home, but they would not go home. They were going to cast tonight. *There's hot metal in that ladle and we're going to put it in the machine.* They begged Millett to let them prove they could cast steel.

"The troops want to go on," said Busse to Millett.

Millett nodded. He gave an order to proceed with a third attempt to cast. A foreman waved his arm and shouted, "Bring in another tundish. We're going to do it."

■ ■ ■

The air in the pulpit had gone high. One aspect of a startup is that people don't change their clothes for days on end. Kolakowski entered the pulpit and said to Owens, "What next? First we had an overflow. Then we had a nozzle break off. What next?"

"It will be perfect this time, Manfred," said Owens. "You see, a new casting machine is like a virgin. It's a lot of work and a tight fit, but it's worth it in the end."

The inventor was not amused. He looked at Owens with a grave expression on his face. Another German remarked to Owens, "I don't know how you know this, Pat."

■

"These casting machines are like women," repeated Owens, dragging on his cigarette. He blew smoke. He removed a hard hat from his head. Its crown was lined with a wadded wet paper towel, a sweat rag. He extracted the sweat rag and put in a fresh wad of dry paper towels, and put the hard hat back on his head. "Look at it this way," he said. "By the time we get a cast, it'll be morning and the bars will be open again. So after we cast we can get a drink and then maybe we can get into a fight." Owens exhaled a breath of smoke, the maroon circles under his eyes deepening by the minute.

■ ■ ■

On the floor of the melt shop, near the casting tower, the casting crew began to plaster the inside of a bathtub with concrete the color and texture of mud. They pressed ceramic boards to the concrete to pack the bathtub with insulation. They worked fast, slapping mud with trowels.

"We'll get it this time. A breakout is the only thing that will stop us now."

"Don't say that word!"

"This is double-barreled, from-the-hip action."

"And you guys thought you'd have a pudge-out today."

Said Kinsey, "How did you like that sound when the mold filled? That kind of *whoosh.*"

"That was pretty entertaining."

"If we couldn't have stopped the pour," said Kinsey, slapping mud in the bathtub, "I'd have probably knocked someone down getting out of there."

It was far past midnight, and morning was coming near. By the door of the melt shop, a knot of vendors and contract engineers smoked cigarettes and discussed the situation in the melt shop.

"These startups are all the same," said a vendor. "I saw six shifts in one mill, before they got a cast. That one took forty-eight hours."

"How long have they been at this one?"

"Since seven this morning."

"That's not so long."

"Remember that one in Mexico? The heat started at seven A.M., and they cast the next morning at five A.M. It'll be like that."

"Yeah, they'll go until sunrise."

"For these guys, this plant has to work. Not because it's a nice idea but because it's their job."

■

228

"Hey, it's a full moon tonight."

The vendors took a moment to admire the moon, their cigarettes huffing in the dark. A waxing gibbous moon glittered over the smooth walls of the melt shop. The melt shop was an insignificant dot of metal on the curved earth, rolling under the moon, and inside the dot about forty organic beings had been attempting to pour a bit of molten iron into a certain shape. The earth is half iron by weight. We live on an ingot that is covered with a thick layer of silicon slag. The moon is all slag. Some of the larger defects on the slag-ball that orbits the ingot have interesting names: the Ocean of Storms, the Lake of Dreams, the Sea of Rains.

"You remember that startup at Harkness Steel?"

A laugh. "Oh, Jesus, I had forgotten Harkness!"

A cigarette glowed.

"Remember that plant manager at Harkness—what was his name?— Buddy Borque."

"Ha! ha! ha! Buddy Borque! Jesus, I had forgotten him. He was so fat."

"God, he was so fat. He was built like a shipping crate with a human head on it. He could have popped a waterbed."

"Are they gonna preheat that tundish or are they just gonna put steel in it?"

"They're runnin' that tundish cold."

"Oh, Jesus! No! That tundish will pop! They gotta preheat it! That tundish will pop!"

"I think they're gonna run that tundish cold."

"I'm positive they will preheat."

"They better preheat!"

The vendors had seen everything. Some of them were not entirely impressed with the Nucor Corporation. One of them later told me that he did not enjoy this Nucor job. It made him nervous. "This startup is not as safe as the ones I've seen at Big Steel companies," he said. "There are too many people on that casting deck. There's no fire equipment or fire personnel up there. The big companies always have firefighting equipment on hand when they start up a casting machine."

"What good is firefighting equipment?" I asked him.

He nodded thoughtfully. "You can't put out a steel fire," he said. "But the firefighting equipment can help to contain the electrical fires that usually follow a big spill. The other thing is, the Big Steel companies take a lot of time training people. Everything is done very carefully. At Big Steel,

■

they spend months running simulations before they pour any steel. Months. I think Nucor is rushing this cast through. And I think they are doing it for two reasons. Number one is, they want to put pressure on the people downstream. They want to get the people downstream to finish their work on the downstream end of the CSP and in the cold mill. And number two, they're doing it for the stockholders. I don't think either of those is a good reason to rush a cast.''

■ ■ ■

People gathered in the pulpit, leaning against the walls and sitting on the floor, to keep warm and to escape the throngs on the casting deck. The air had died inside the pulpit—the smell could have crashed a computer. The floor was a tortured mass of cigarette butts. Owens, the commissioning engineer, had been reduced to a cauterized nerve-stump. This happened to him every time he started a casting machine. His face was covered with a thin film of sweat. He was as physically organized as a chopped clam. He begged a Camel Filter from somebody and attempted to light it with his Scripto. His hand swerved, the flame hit the middle of the cigarette, and the middle of the cigarette caught fire. He drew the lighter along the whole cigarette, it burst into a conflagration, and he inhaled smoke into the last abyss of his lungs.

"You look terrible, Pat. Nothing personal," said a steelworker.

"No problem," said Owens. "We have the technology."

On the deck, Mark Millett was talking to Rodney Mott, the soft-spoken manager of the hot mill. "It looks like we're going to watch the sun rise," said Millett.

"I hope it works this time," said Mott. "There comes a point when extra human effort won't overcome a machinery failure."

Keith Busse came over to his mill managers to chat with them. "They were having all kinds of trouble in the cold mill today," he remarked. "I'll tell you what I want to do. I want to wire some *plastique* explosive to the cold mill, attach all kinds of wires to it, and push the plunger."

"I think somebody already did that to the cold mill," said Rodney Mott.

"I already have a feeling," said Busse, "that we are going to be hard-pressed to get a million tons a year out of this caster. Look at what happened tonight. To think that they're going to have to turn this machine around in a few minutes." Busse put his hands in his pockets and stood so close to Mark Millett that his stomach nearly shoved into Millett. "Mark—

■

what's wrong with modifying this deck to put a second caster in, right over there?'' Busse nodded toward the east end of the deck.

Millett grinned. They strolled over to the east railing, where they looked across a huge empty bay at the east end of the melt shop.

"You might have spent a little more money adding this extra space, Mark, but you have greater freedom to put in a second caster if you want it,'' continued Busse.

Millett nodded and said nothing.

Busse spotted a chromed bolt lying on the deck. He picked it up. It was shiny, five inches long, and weighed half a pound. "That's a Nucor bolt,'' he said, turning it over in his hand. "You can tell by the head-mark. See? It says *n*. That's a good-looking bolt. I'm proud of that bolt. We're not making any money on bolts. This bolt was made by a not-for-profit organization, but it was made well.'' A smile flickered over his face, but his green eyes were hard.

Meanwhile, over at the metallurgy deck, the steel had been sitting so long in the ladle that the metallurgy crew began to worry that it might eat through the ladle's lining. Then the ladle would burst like a wet paper bag. So they poured the steel into a fresh ladle, juggling it from one ladle to the other. Then they reheated the steel with the arcs at the metallurgy deck.

At 4:10 in the morning, the last bathtub was red-hot and ready to accept hot metal. The bathtub was fitted into place over the mouth of the casting machine. A siren went off, signaling the movement of the ladle back toward the casting deck. The casting crew put on their silvers again.

People had been trickling up the stairs of the casting tower all night, determined to stay on the casting deck this time, come hell or high steel. Faces in the crowd were familiar. Ken Kinsey remarked, "What could be better than four o'clock in the morning to run your first slab?'' Dick Haltom, whose job it was to take the temperature of the steel, went home, having been awake for forty-eight hours. The scrapyard supervisor, Jim Hoskins, volunteered to take the temperature of the steel. Manfred Kolakowski paced in a corner. Patrick Owens emerged from his fetid lair and remarked to Kolakowski, "You look tired, Manfred.'' Kolakowski shook his head: Not tired. James McCaskill, wearing a green furnace suit, loitered at a rail, smoking a cigarette under his twitchy mustache. Jan Roach, the safety director, pushed through the crowd. "Get everybody off the deck!'' she shouted. Nobody moved. Mark Millett took off his hard hat and ran his hand through his graying hair. "This is what Nucor is all about,''

■

he said. Keith Busse stood alone at the rail with his back to the crowd, watching the ladle of steel approach the casting deck.

The crane placed the ladle on the ladle car, and the car moved the ladle slowly over the casting deck until it came to a halt over the bathtub. The ladle was absolutely huge. It blocked the glow of lights shining on the ceiling of the melt shop, throwing the steelworkers into shadow. It contained an immense amount of liquid steel.

Kinsey and two other steelworkers knocked fists and elbows together, gave a violent shout, took up their stations around the machine, and crouched lightly on knees that were slightly bent.

The situation was no longer under any manager's control—the steelworkers had taken control of the deck. To me it evoked a certain moment in the *Iliad,* when the Achaeans, who have been trying to take the city of Troy for nine years, almost decide to give up and go home. But a dangerous spirit enters their minds and whispers in their heads until they are drunk for war: "Now warfare seemed lovelier than return, lovelier than sailing in the decked ships to their own native land."

Kinsey shouted to his commander, Thompson, "Test open?"

"Yeah," said Thompson. "Do a test. Cast mode!" he barked, turning 360 degrees slowly.

The sequencing began.

Manfred Kolakowski approached the bathtub and took the injection lever in his silvered hands. This time, Kolakowski would feed the machine. Kolakowski gingerly moved the injection lever up and down, by about two inches, in a dry test. He now understood two things: (1) That the lever was almost impossible to control, and (2) That any movement in the lever greater than two inches would trigger a breakout, a metal gusher spewing across the deck or pouring down inside the machine.

"Everybody ready?" said Thompson. "Let's go. Open slidegate."

The slidegate opened full stroke and a heat wave rolled over the casting deck as steel shot full-bore into the bathtub. The bathtub filled quickly, while Miller cried out the weight of the steel in the bathtub at brief intervals: "Forty-three hundred pounds! 5,020! 5,940! 6,120 pounds! . . ."

Hoskins, the scrapyard supervisor, climbed onto the bathtub car and stared into the rising bath, holding a temperature probe in his hand.

"Fifteen thousand!"

"Temperature!" said Thompson, and Hoskins slid the probe into the bath. The temperature came up on the display board: 2,778 degrees.

■

Thompson saw that the steel was a bit too cold for a good cast. But he was not going to abort this one. Since he was not going to abort it, "Open stopper!" he shouted to Kolakowski.

Kolakowski pushed downward on the injection lever to open the bathtub stopper, but nothing happened. Kolakowski heaved again, but still nothing happened.

Once again, the stopper had been frozen by a clot of skull. Meanwhile the bathtub was filling with steel. An air of crisis swept the deck.

Kinsey raced over to help Kolakowski. The steelworker and the inventor struggled with the lever, Kinsey shouting, "It won't open!"

"Twenty-one thousand!" cried the foreman.

"God damn it!" shouted someone else.

"Twenty-five thousand! Twenty-seven thousand!"

The pour from the slidegate continued at full throttle. A spray of ejecta billowed from the bathtub and splattered over the deck, and the steelworkers ducked.

"Twenty-eight thousand!"

The bathtub was nearing overflow.

"Thirty thousand pounds!"

"God damn it!"

The voices of the steelworkers rose in a tense hysteria.

Thompson already knew that somebody was going to have to burn open the bathtub's drain hole with an oxygen lance. That was the only way. Someone was going to have to put an oxygen lance into that naked bath. Thompson swung around on his feet and cried, "Lance it!"

"Thirty-two thousand!"

The bathtub was within thirty seconds of overflow.

"Burn it!" Thompson roared.

Big Dave Thompson looked around. No one moved. No one knew who was supposed to lance the bath. Thompson grabbed an oxygen harpoon. He was not wearing a faceshield. No time to find a faceshield. He cracked the lance valve and oxygen hissed from the tip of the lance. Thompson jumped onto the bathtub. He drove the lance into hot metal, fishing the jet through liquid steel, feeling for the drain opening of the bathtub. He felt the lance strike the stopper. Keeping the tip of the lance against the stopper, he guided the tip down into the drain hole. There was a thud in the bathtub. A terrifying spray of sparks billowed into the air, and Thompson vanished in the spray.

The sparks covered Manfred Kolakowski. The inventor's body seemed

■

to lurch up through fire, and he threw his body over the lever. The lever jerked down and steel flooded into the machine.

"Come on, baby!" someone yelled.

The mold glowed and brightened and the copper horn sounded a deep note. A spray of sparks flew from the mold. The crew ducked and dodged sparks. Thompson was still lancing the drain hole. Sparks waved out of the bathtub, and they looked like nodding wheat.

"*Powder!*" came the voice of Thompson.

Kinsey began to push powder into the hot mold. Suddenly Busse was next to Kinsey, pushing powder, too. The vice-president and the steelworker crouched in the glow, sparks racing around their legs.

Thompson jumped down from the bathtub, tossing away his glowing lance. "*Strand start!*" he roared.

The mold shook up and down, faster and faster. Sparks sailed from the mold and bounced across the deck like ping-pong balls as a ribbon of steel moved downward into the machine; the cast had begun. A dribble of steel gurgled across the shaking mold table and solidified into a skull. Kinsey dropped his powder hoe and bent over and ripped the skull from the table with his gloved hands. Busse pushed powder into the mold. Kolakowski twitched the lever. A roar came up from the crowd below the casting tower. They had seen steel begin to emerge from the machine.

Suddenly the machine ground to a halt, with a sound like *wah, wah, wah, wuuuh*.

"What happened?"

"It stopped!"

"Breakout! *Breakout!*"

There was a thud—a steam explosion under the deck. Smoke and flames burst up around the mold cover and flickered and flared in the mold. Steel drained out of the Kolakowski mold, raining and splattering down through the casting tower. The roar from the crowd below turned into angry bitter yells when people saw the hot metal cascading down through the tower.

Kolakowski yanked up the lever to shut off the flow into the machine. The slidegate was shut. "Let's get the ladle out of here!" exclaimed Millett, and the crane snatched the ladle away from the casting deck.

Suddenly the steelworkers abandoned their posts. They dashed toward the exit stairs, tearing off their silvers as they ran, throwing their gloves into the air. They made it to the base of the casting tower in seventeen seconds.

There, amid a welter of machinery, a tongue of cherry-red steel poked

■

from the machine, wreathed in clouds of steam. It was a thin slab, two inches thick and as wide as a sidewalk. It protruded six feet from the base of the casting tower.

"By God, the son of a bitch works!" said a steelworker. "All we got to do now is get the bugs out of it."

A lot of handshaking took place. "Thanks for your work, gents," said Millett to his men.

"Mark, you are coming away with a good one for Nucor," said Thompson.

The casting crew wanted to try again for a fourth attempt. "Come on, Mark. How about it?"

"Nope, mates," he said.

"Hell, I'm disappointed," said Kinsey. "We made a piece of steel six feet long. I wanted to see a *slab* come out of that Tinkertoy."

"Yeah, but it's a world record in the western hemisphere," remarked Owens.

Kolakowski seemed disappointed. "It has been a long day," he said, pulling off his silver gloves. "A long day and a long job, to get two meters of slab. But a lot of questions have been answered. We will cast again on Monday."

Keith Busse said to me quietly, "Mark Millett learned a lot about leadership tonight. The men were begging him to go on. Do you remember General George Patton at the Battle of the Bulge, when Patton said, 'I can be there'? Mark got there. That's the lesson Mark learned tonight."

Busse jumped in his car and drove home as fast as he could, calling David Aycock on his cellular telephone. He woke Aycock out of bed to tell him that Crawfordsville had made six feet of steel.

I walked out of the melt shop into the parking lot to see what was happening in Indiana besides steel. It was a cool Saturday morning, a few minutes before sunrise. Nothing obvious was happening in Indiana. Sparrows dropped into the shaved wheat around the mill buildings. A squall line traced the eastern horizon with a ridge of blue bumps. A stiff wind began to blow and abruptly the sun cracked the blue bumps and kept on rising. The day smelled like summer.

■

16

ROLLING MILL

■ ■ ■

hree days later, they tried again. Manfred Kolakowski again fed the machine by hand. He stood with one foot on the machine, and the machine heaved up and down, shaking Kolakowski to his bones.

"Go, go, go, go!" the steelworkers chanted at Kolakowski.

His face poured with sweat. His gravid jaw hung loose. His faceshield chattered around until he could hardly see what he was doing. The machine repeatedly threatened to break out, and each time, Kolakowski moved the lever by half an inch, up a little, down a little, against heavy pressure. It took all his strength. He hung on for half an hour, while a slab emerged slowly from the machine.

"A hundred feet, Manfred!"

"Two hundred feet!"

The cast ended in a breakout, when some overexcited person in the pulpit bumped against a chart recorder, and the machine got spooked and poured hot metal down through itself with a rush of pale flames. Kolakowski staggered away from the lever, through the smoke of the breakout.

He had produced a long thin slab of steel, 2 inches thick, 52 inches wide, and extending 285 feet into the tunnel furnace, a piece of steel with the proportions of a fettuccina. When the slab had cooled, steelworkers cut it into lengths with torches, a crane laid the lengths on the ground, and the inventor crawled over every inch of the steel on his hands and knees, brushing it with a wire brush, looking for cracks. Every time he found a crack he noted it in a notebook.

■

"You go like a cat over the steel," as he put it.

He found plenty of cracks.

No matter, that evening the entire melt shop went blind drunk at the Scoreboard Lounge. The bash at the Scoreboard slowed the Crawfordsville Project for a couple of days, but after an auction of Buicks and some rest, the melt shop pulled itself together for a third cast. This one was a fiasco, an immediate breakout. A fourth cast produced a long slab, ending with a violent breakout. At each attempt, Kolakowski took the injection lever and worked it until he nearly dropped, but gradually the machine logic took control of the lever. On June 28, Mark Millett celebrated his thirtieth birthday.

Cast number six extended five hundred feet through the tunnel furnace and almost touched the millstands of the Compact Strip Production machine. At that point Millett had to shut down the casting machine, because the steel had nowhere to go—the CSP's millstands weren't ready for steel.

Millett had felt a monkey sitting on his back. As he put it, "When that slab went down the tunnel furnace the dreaded monkey was sitting on it. The monkey rode the slab down to the end of the tunnel furnace and jumped on Rodney Mott's back."

By July 13, Mark Millett had attempted nine casts. Of those, four were aborted and five were considered to be successes, even though all but one ended with a breakout. After each cast, Kolakowski crawled the length of the slab on his hands and knees, brushing the surface with his wire brush and taking notes.

Now the pressure was on Rodney Mott, the hot mill manager, to try to drive a slab through the millstands, to crush it flat between the rolls into sheet steel and coil it up in the downcoiler, at the end of the line. In late July, Mott told Busse that he was ready to drive a slab into the millstands for the first time. He assured Busse that the first attempt to roll steel would trigger a millwreck, a jam-up of red-hot steel inside the millstands.

"All right. And we'll try again," Busse replied.

On July 22, a crowd gathered inside the hot mill building, two hundred yards from the casting machine, to watch the first slab be driven into the millstands. The casting machine produced a long slab that traveled through the tunnel furnace, emerged from the tunnel furnace, and went through all four millstands—*ba-boom, ba-boom, ba-boom, ba-boom*—being wrung thinner and thinner between iron rollers, like dough being rolled flat. The rolling machines, driven by four 10,000-horsepower Westinghouse electric motors, squeezed the slab down from 2 inches to $\frac{1}{10}$ of an inch, and it emerged from the millstands as a sheet of red-hot steel, 52 inches wide and as thick as cardboard. A hundred feet of the sheet steel emerged from the millstands and tore away like a broken noodle, and headed down the line for the downcoiler. The crowd erupted in cheers and ran alongside the CSP, following the metal as it moved along. Busse ran with the crowd, shouting his lungs out. The noodle nosed into the downcoiler machine, and the downcoiler wrapped it up into a coil with a snap. Meanwhile, back upstream in the rolling mill, the rest of the steel jammed inside the millstands.

A hundred feet of torn steel had made it into the downcoiler—a pathetic little coil of Distressed Material weighing not more than one ton. It was the Crawfordsville Project's first ton of finished hot band steel. It had been made with three million man-hours of labor, the human labor needed to construct the Crawfordsville Project. Somehow Keith Busse had to ramp the mill's productivity up to one man-hour per ton. It did not seem possible.

■ ■ ■

It was a long, mean summer. Weeks in August went by when Mark Millett would send Rodney Mott a slab of steel, and Mott would drive the slab into the rolling mill, causing a wreck. When the steel didn't wreck, a coil of Distressed Material rumbled into the downcoiler. The DM went back into the furnaces.

One day there was a terrible millwreck, when steel jammed in great loops inside the millstands. Rodney Mott went down to the floor to see what he could do to fix the machinery. He climbed inside a millstand, carrying a wrench. The steelworkers could hear their manager banging around inside the machinery—they had no idea what he was doing—and finally Mott emerged covered with grease. "I think I fixed it," he said. "There are only two possibilities. Either the next slab of steel will go through the rolls perfectly or it'll wreck worse than before." They restarted the CSP, sent a slab into it, and the metal was rolled perfectly, that time.

Mott had got his hands dirty inside a machine. Managers at Big Steel companies do not get their hands dirty inside machines. There are two reasons for this. In the first place, Big Steel managers do not have any authority to do solo tinkering. Solo tinkering is absolutely forbidden. You could lose your job if you tried it. A manager has to get permission from the chain of authority before he can experiment with any machine-settings that could cause a millwreck. In the second place, union work rules won't allow managers to touch a machine during routine operation. Only unionized steelworkers may touch a machine. Not even foremen are allowed to touch machines at Big Steel. If a foreman touches a machine at a Big Steel

■

company he can have a union grievance filed against him. (Although Dave Thompson, who had been a foreman at Rouge Steel, explained, ''We did anyway. We weren't supposed to touch the machines, but we did. If you did it too much, the union might file a grievance against you.'')

The mechanical problems at Crawfordsville seemed to get worse, not better, as time went by. The furnace crew would fire up the furnaces at dawn, trying to get a heat of steel, and finally, around ten o'clock at night, the casting crew would have the casting machine in a state of readiness. They'd pour a heat into the caster, a slab would come out of it, and the slab would head down the line through the CSP's tunnel furnace. But when the slab hit the millstands and the Westinghouse motors kicked in, the immense draw would kill the electric power throughout the entire Crawfordsville Project. It happened again and again: a slab would hit the rolling mill and the lights would go out. Then the furnaces would die and the casting machine would break out, steelworkers shouting, ''Breakout! Breakout!'' Meanwhile, every computer at the steel mill would crash, and metal would jam inside the rolling mill. Meanwhile, Nucor's customers began to place orders for steel. They began to call Nucor's salespeople, asking, ''So do you think maybe you can send us a little steel?''

■ ■ ■

Keith Busse rattled around in his own cage that summer, worried about the business question, the question of profit. The steel mill was now producing small quantities of the two types of sheet steel: hot-rolled steel and cold-rolled steel. Hot-rolled steel, known in the trade as hot band, was the sole product of the Compact Strip Production machine. Hot band or hot-rolled steel is about one-tenth to one-quarter of an inch thick, and has a rough blue-black surface, or it has a silvery surface if it has been pickled in acid and oiled. Cold-rolled steel, by contrast, is smooth, very thin (rolled down to a few hundredths of an inch thick), and it can be painted or enameled. Some of the hot band steel that came out of the CSP would be sent to the cold mill, to be thinned and elongated, tempered, and polished—that is, to be made into cold-rolled steel.

Hot band steel is for heavy and sometimes crude applications. The frame of an automobile is made of hot band. Hot band is made into wheel rims, automobile bumper stock, dumpsters, garbage trucks, door hinges, pipe brackets, water towers, earthmoving equipment, railroad cars, truck beds, the guts of washing machines and dryers, bed frames (most everyone sleeps on hot band steel), shovels, electrical outlet boxes, harvesting machines, tractors, back-end loaders, and architectural railings. When rolled up into a tube, hot band steel becomes pipe.

Cold-rolled steel is made into automobile hoods, doors, fenders, into the exterior shells of washing machines, dryers, hot-water heaters, refrigerators, and microwave ovens. Cold-rolled steel goes into the guts of computers and VCRs. A lot of cold-rolled steel goes into offices in the form of filing cabinets, lighting fixtures, hung ceiling brackets, radiators, and vents.

Steel is cheaper than chicken parts. Hot band steel ordinarily sells for between $340 and $440 a ton, between 17 and 22 cents a pound. Busse hoped that the CSP would manufacture hot band steel at a cost of about $250 per ton—12½ cents a pound. That would make Nucor the world's low-cost producer of sheet steel. Twelve and a half cents a pound was counting everything—labor, energy, raw materials, depreciation, taxes, insurance, and corporate overhead. Busse believed that the most efficient integrated steel companies in the United States enjoyed, at best, a manufacturing cost of $310 a ton, 15 or 16 cents a pound. That was if you counted all of Big Steel's costs, including insurance, taxes, depreciation, and corporate overhead. As for the Asian and Latin American producers, Busse believed that the cost of shipping steel across an ocean offset the cheaper costs of the foreign steelmakers. As he put it, "I don't think the Japanese can land a coil of hot-rolled steel here for less than three hundred and fifty dollars a ton. And I doubt that the Koreans can do it for less than three hundred and ten. And the truth is that the South Americans are not very efficient producers."

If the CSP could make steel for $250 a ton, then Keith Busse would have perhaps a fifty-dollar-a-ton cost advantage over the world's most efficient steel producers, a cost advantage of 2½ cents per pound. It seemed incredible, but there it was, staring Busse in the face: American steel might be the world's low-cost steel. This could be the biggest turnaround since Harry Houdini escaped from a coffin at the bottom of a frozen river. Keith Busse might be able to undersell everyone in the American Midwest. His net after-tax profit ought to come in at around 4 or even 5 cents a pound, depending on the price of steel. A nickel a pound in profit.

Four or five cents a pound in profit didn't sound like much, but it added up to $80 or $100 a ton in profit. A net of $80 on a ton, times 1 million tons a year, could add up to $80 million in annual profit for the Nucor Corporation. The Nucor Corporation had earned $70 million in 1988, an exceptionally good year. When Crawfordsville came on line, Keith Busse might be able to more than double his corporation's profits. The arithmetic stared him in the face. It also stared Wall Street in the face, because Nucor's stock was trading high, on high hopes. If the profits didn't materialize, Wall Street would take care of the stock. The pressure on Keith Busse was excruciating. It was the pressure of a mountain of nickels.

■ ■ ■

The aura of triumph that had filled the Crawfordsville Project when Mark Millett and Rodney Mott began to run steel all the way through the CSP gave way to a deepening sense of fear. It began to dawn on everyone that to make a few sheets of DM is not the same thing as to make a million tons of steel and sell it at a profit. By early September, the Crawfordsville Project was still losing $1 million in cash every week, and Distressed Material was piling up in the downcoiler.

Keith Earl Busse couldn't help but notice the success of the Blytheville Project in Arkansas, the Nucor-Yamato I-beam mill, which was now bringing home sharp profits. John D. Correnti, the general manager of the Blytheville steel mill, had done well. The competition between Correnti and Busse, of the rhyming last names, was real and could not be denied. It didn't help that Correnti was minting money while Busse was losing money like it was going out of style.

Every module, every subsystem, every mechanical device in the CSP failed at one time or another, often simultaneously. On one occasion, a slab of steel went crooked inside the tunnel furnace and tore out a wall of the tunnel furnace. Water systems inside the tunnel furnace overheated and exploded. The tunnel furnace boomed with steam explosions up and down its length, until finally the tunnel furnace had to be broken out and rebuilt; the whole tunnel furnace was a tear-out. The casting machine experienced breakout after breakout—"A breakout a day," as Mark Millett phrased it. Virtually every attempt to cast steel produced a couple hundred feet of slab cut short by a nasty breakout. They'd be casting steel, everything going along fine, and suddenly flames would lick up around the deck, to shouts of: "Terminate the cast! The sucker broke out."

"Some of us even wondered if the process would work," Busse admitted.

■ ■ ■

On Wall Street that summer, the money managers had been bidding Nucor's stock higher and higher, fighting over a tight supply, but by September, Nucor was still pouring $1 million a week into the Crawfordsville Project, and the money managers began to get nervous. It began to dawn on Wall Street that Nucor was going to report sharply lower earnings for the September quarter. The corporation's cash was circling a drain in Indiana.

Money managers run investment portfolios for pension funds, mutual funds, and insurance companies, and they control most of the common stock in the United States, buying and selling it in blocks. When they're scared they can hurry away from a stock the way a cat is driven from food by a handclap. They simultaneously began to unload Nucor in unobvious little blocks, trying to sneak it out the door in chunks *before* Nucor reported lower earnings. They call it profit-taking on Wall Street, which generally means that a money manager who paid too much for a stock is blowing it off at a loss.

Nucor traded at $67 a share before Labor Day, and then it had a Labor Day sale. Blocks of Nucor began to pile up on the sell side of the order book of the Nucor specialist on the trading floor of the New York Stock Exchange. Some smart money was shorting Nucor, and the stock had a nice little breakout. It cascaded at a rate of nearly a point a day. The money managers, who a month before had been anxious to buy Nucor, were now anxious to get Nucor out of their portfolios, but they had to drop their asking price a half a point each time they sent a wheelbarrow of Nucor up to the specialist, just to smoke out a bid for the merch from some other money manager. The stock was trading on down-ticks, ticking down an eighth, a quarter, a half, on substantial volume. On September 15, some large blocks of Nucor crossed the ticker tape at around 61, the stock having lost 6 points in as many trading days. Iverson started getting telephone calls from nervous financial analysts.

All they had to do was pick up the telephone. "Hello, Ken Iverson! Yeah! I'm free right now. Sure." He tried to explain to them that it was normal for a startup of a steel mill to take longer than he had figured. "In Crawfordsville, Indiana, we have one of the larger research units in the world," he told them. "It's a one-million-square-foot laboratory. We have two hundred and seventy million dollars tied up in process research." He felt obliged to mention that the plant was still losing one million dollars a week.

The analysts hung up the phone feeling queasy. It was their job to give information and advice to money managers. They called the money managers to report that Ken Iverson had said that the Indiana steel mill was a laboratory that was losing a million dollars a week. When the money managers heard the word *laboratory* it made them feel sick. It reminded them of those college chemistry classes they all got D's in, which is why

■

they had had to become money managers. They called up their brokers and told them to sell the Nucor out of all the portfolios, quick.

Busse thought that Iverson should have kept his mouth shut. "Ken ought not to warsh his shorts in public." As for the analysts, Busse was beside himself with indignation. "They don't understand a startup! They ought not to speculate in stocks! They're so wishy-washy! They're such opportunists! I wish they'd go hide in a hole in the ground!"

For every seller there has to be a buyer. Someone was buying Nucor, but the buyers coolly cut their bids for the stock, and so the stock came downstairs delicately, like a drunk fighting for balance, hoping that no one will notice. Toward the end of the month the stock touched 55, and then recovered a little bit.

Every time Mark Millett opened his newspaper to the stock quotes—he owned Nucor stock—he saw that shareholder wealth had been vaporized. The problem to Wall Street was the casting machine, and therefore the problem was Mark Millett.

During those weeks in September the total stock-market value of the Nucor Corporation withered by $252 million. That was just about equal to the capital cost of the Crawfordsville Project. It was as if the stock market had decided to discount the cost of the Project from Nucor's stock price and pretend that the Project had never happened. Mark Millett began to wonder how much of it was his fault.

During one week in September, Millett made ten attempts to cast steel and had ten consecutive breakouts.

"We were zero for ten that week," as Keith Busse put it.

The following week the CSP produced its first prime coil of steel, a strip of steel without any serious blemishes.

■ ■ ■

On a warm, sunny October morning, at the height of Indian summer, Ken Iverson sat on a chair in Keith Busse's office at the Admin Building, lacing up a pair of steel-toed boots. This was Iverson's first trip to the Crawfordsville steel mill since the startup had begun. Iverson wanted to see what was going on.

He tucked pin-striped trousers into the boots and tied the bootlaces. He wore a wine-red silk tie and a blue pinpoint cotton shirt with the monogram FKI on the pocket. Sunlight fell through a window and over Iverson and gleamed on the chrome bolt on Busse's coffee table. The mill had not yet shipped a coil of steel to outside customers.

■

"I'm not worried," said Iverson. "I've been telling people I sleep like a baby. I wake up every hour and cry."

He wasn't kidding about waking up. Iverson slept poorly during start-ups. "I don't really even try to get back to sleep," was the way he put it.

Keith Busse sprawled on a chair facing Iverson, snapping a coffee stirrer between his teeth. He wore a plaid shirt and stonewashed Levi's. He held the coffee stirrer away from his mouth and took a sip of coffee from a Styrofoam cup.

"Did you hear this one, Ken?" said Busse. "The U.S. Steel people called a huge meeting at the Lake Shore with their vendors. They put five hundred vendors in a room, and this big shot of a vice-president got up and said U.S. Steel would not tolerate it if the vendors do business with Nucor for less money than with U.S. Steel."

"No kidding!" A grin crumpled Iverson's face and stretched his ears. "Where'd you hear that?"

"One of our vendors. He was at the meeting. He thought it was comical. U.S. Steel is big-time scared of us, Ken. What they're doing is threatening their vendors—if that ain't being scared to death of Nucor I don't know what is. Those U.S. Steel people must have some reservations about their ability to negotiate or they wouldn't bully their suppliers like this."

"U.S. Steel shouldn't be scared of us. A million tons is nothing. They shouldn't even know we're here."

"I also think they're talking up the Nucor Threat for their own ulterior motives," said Busse, thoughtfully. "So they can deal with the unions."

"I've heard that U.S. Steel may build one of these Compact Strip plants."

"I doubt it." Busse slurped his coffee. "They're risk-averters, not risk-takers."

"Big Steel will survive," said Iverson. "They'll survive! Even if they have to bust it all apart."

A telephone rang on Busse's desk. Busse crossed the room and yanked up the receiver. "Yeah, Mark. What kind of a jury rig is it? . . . Okay, if the jury rig works, let it alone until Sunday." He hung up and turned to Iverson. "They're finishing the heat."

Iverson stood up. He pulled a green spark-proof jacket over his starched shirt and fastened the jacket's snaps. He put a white hard hat on his head. He wiped a pair of safety glasses with a handkerchief and put them on. He took a sip from a Styrofoam cup of coffee and left it on the table beside Busse's chrome bolt. He and Busse went outdoors into a

■

dreamy October morning. In the distance, belts of unharvested corn were drying on the stalk. The two executives headed for the melt shop, a tan steel building that loomed above their tiny figures, and they found Mark Millett in his office.

Millett was in a quiet mood. Nobody shook hands. They passed through the repair shop. Scattered all over the floor stood blackened pieces of the casting machine, scorched and slobbered with frozen metal. Two guys on ladders were cleaning up a module with a chisel and a cutting torch. There was a blue sparkle and a hiss, and black flakes of steel fell to the floor. The steelworkers grinned at Iverson.

"Hi!" said Iverson.

"Hi!"

"Hi!"

On the casting deck, a casting crew was preparing to cast. The crew moved a bathtub over the mouth of the machine, and lowered the bathtub until the nozzle protruded down into the Kolakowski mold.

A siren went off, and the ladle, dangling from an overhead crane, approached the casting deck. The casting crew began to hurry here and there with practiced movements; one man got down on his belly and fitted two small pieces of wood into the mold.

"There's an element of basic fun in making steel," Iverson remarked suddenly to me. "They like to do it. That's the way hot metal men are."

The ladle creaked over the deck and stopped over the bathtub. A foreman gave a hand signal and suddenly a bright pour erupted through the ladle slidegate into the bathtub.

Iverson crossed the deck to get closer to the hot metal, with his hands in his pockets.

"Twenty-two thousand!" shouted a steelworker.

Patrick Owens, in the pulpit, remarked, "Some big bosses are here today," and he hit a button, the nozzle opened, and steel thundered into the Kolakowski mold. Sparks skittered out of the machine and around Iverson's boots and trousers, and the *whooshing* sound of the copper horn echoed over the deck.

The foreman counted down—one! two! three! A steelworker started hoeing powder into the mold, the machine heaved up and down, and the cast began. Iverson moved closer to the mouth of the machine. He bent over until his face was inches from molten steel. Flames flashed from the mold and Iverson became a shadow in the flames, just another hot metal man peering down the throat of Iverson's machine.

■

Iverson followed the flow of steel downstream. He descended the tower's stairs with Keith Busse, leaving Mark Millett on the deck. Iverson was constantly in motion inside a steel mill, and he often moved faster than the steel itself. At the base of the casting tower, a glowing thin slab of steel, the color of a tangerine, was moving out of the casting machine and entering the tunnel furnace. Heat waves wriggled off the steel. The warmth was colossal. There was a shearing machine below the casting tower. It sheared the steel coming out of the tower into one-hundred-and-fifty-foot lengths.

Iverson and Busse headed for the CSP's millstands, walking northward along the tunnel furnace at a rapid clip, hoping to beat the slab into the hot mill. Five hundred and forty feet down the line, they entered the hot mill building. There, the CSP's four millstands towered toward the roof, under arrays of powerful sodium lights in the roof. The ceiling lights converged in lines toward infinity, toward the lower end of the building, where the downcoiler waited to accept a sheet of hot band steel.

Iverson walked past the millstands with his hands in his pockets, on long casual legs, inspecting the millstands.

Each millstand consisted of two tall doughnut rings of forged steel, the so-called roll housings. The holes of the doughnuts contained bearings. The bearings held rollers. The rollers—more properly known as rolls— were made of polished cast iron, and they resembled rolling pins. They rolled the steel flat. Two opposing work rolls bit the steel like the wringers in an old-fashioned washing machine. Two massive backup rolls pressed against the work rolls from above and below, exerting tremendous pressure on the work rolls and therefore on the soft, red-hot steel as it passed between the rolls. The four rolls in each millstand weighed a total of 175,000 pounds—almost 90 tons. The weight of the rolls alone wasn't enough to crush red-hot steel, and so the rolls were boosted by hydraulic rams. The rolls turned around and around, attached to drive shafts. The drive shafts led into a building beside the hot mill, which contained the four 10,000-horsepower Westinghouse electric motors.

Iverson remarked, "The drive shafts that drive these mills are three feet in diameter."

Iverson led the way up a flight of stairs to the hot mill's control pulpit.

■

Inside the pulpit, Rodney Mott, the hot mill manager, was bent over a computer screen, studying the parameters of the machine.

The pulpit was dark inside, and it glowed with a half-dozen color video displays and rows upon rows of buttons, switches, and joysticks—the command deck of a starship. A room nearby was stuffed with mincomputers, and the minicomputers contained cyberspace. The cyberspace contained the three levels of software that were needed to run the CSP. Some of the software was running and some of it hadn't been written yet. The CSP had not yet evolved into a robot.

"How's it going?" said Iverson to Mott.

"We're doing all right," said Mott.

A pulpit operator said, "Slab's gonna be here in a minute."

"We'll watch it from down below," said Iverson. He and Busse descended to the floor of the mill and stood near the millstands to watch the steel come through.

Rodney Mott hovered beside a pulpit operator, at the control desk of the starship. Mott was a little bit nervous. He wanted to show the chairman a nice smooth operation, and so he figured he'd let the operator work some of the joysticks himself, on manual, rather than letting the computers do it.

Now the nose of a glowing slab—the leading edge of the cast—emerged from the exit door of the tunnel furnace and headed slowly for the millstands. The slab was 2 inches thick, 52 inches wide, and 150 feet long, having been sheared to that length as it emerged from the casting tower. It nosed into the first millstand. The rolls bit the steel. They grabbed the metal. When the slab's nose hit the first roll bite, it gave off a *ba-boom!*— the sound of red-hot steel nosing into a roll bite. The slab was squeezed through the roll bite, emerging half as thick and moving twice as fast. Water jets drenched the rolls to keep them from overheating. The red ribbon hit millstand No. 2—*ba-boom!* It came out thinner and moving faster. It hit millstand No. 3—*ba-boom!* It came out fast, thin, and fluttering. It wriggled. It hit the last millstand, *ba-boom!* It emerged as a red-hot sheet of plain carbon steel a tenth of an inch thick, moving at 10 miles an hour, traveling down the line.

Busse said, "When you think of taking a one-hundred-and-fifty-foot slab of red-hot steel, and elongating it into a band that's two or three thousand feet long, and it doesn't change its *width* by even a quarter of an inch when you stretch it like that, it's just incredible. You take a piece of

■

taffy and stretch it, and the center of the taffy gets narrower. Why doesn't that happen with steel? Steel is incredible stuff.''

In the pulpit, Mott didn't like the way the steel was wriggling.

The machine operator, a young guy with longish curly hair poking out from beneath a hard hat, muttered, ''Damn it, damn it, damn it!'' while he toggled a joystick.

There was something wrong. As the steel moved from the third to the fourth millstand, the steel humped up in a fluttering loop. The steel developed a small tear that widened into a rip, and the steel talked in a weird voice—*woogah, ba-ba-ba!*

Now Mott started to toggle a joystick—no, no, not a millwreck in front of the chairman. His finger hovered over an abort button.

''Not yet!'' snapped the operator to his boss, Mott. ''It's gonna go through!''

Mott punched the abort button. ''Too late,'' said Mott.

The steel came to a halt, jamming between the rolls. It was a millwreck in front of the chairman.

''Let's shear it and cut it out of there,'' said Mott, over his shoulder. ''Call the caster and tell them slow down the cast.''

''They've terminated the cast!'' cried an operator.

The chairman, down on the floor, bent over and peered into the millstands at a glowing ribbon of steel looped into an S—a mangled strip of steel, also known as a cobble.

Busse hurried upstairs to the pulpit. ''What happened?''

''It got too loose,'' said Mott. ''Then it started folding and tearing.''

Down on the floor, a steelworker with a cutting torch began to slice the mangled steel out of the millstands.

''It's all part of the problem of running these machines tied together,'' Iverson remarked. ''The steel snapped. Then they had to stop casting.''

Busse peered into the millwreck. ''A video-game player would be a damned good mill operator,'' he said.

''It could take two years for inexperienced people to get really good and smooth at this,'' said Iverson.

The steelworker finished slicing the cobble. He hooked a chain to it, and an overhead crane dragged the cobble out of the millstands and dropped it on the floor with a clank. It was a wrinkled, folded hunk of gunmetal-blue lasagna.

Busse stepped on the metal and bent over, gazing at its surface. ''The whole thing cobbled like hell,'' he muttered.

■

"That first millstand is really sluggy today," said Iverson, kneeling next to Busse. "You could tell from the sound of it." Iverson removed his safety glasses, hooked the earpiece between his teeth, and squinted at the folded metal. "The edges look good," he said. "The surface looks terrible."

Iverson and Busse continued their walk downstream. Below the millstands there was a runout table, a set of rollers 390 feet long, along which the steel travels while water jets cool the steel. Iverson and Busse arrived at the CSP's downcoiler, a complex machine that sat in a concrete pit. Nothing was happening at the downcoiler. Beyond the downcoiler there was an elevator ramp, 210 feet long, that delivered coils of steel to an inventory area. The end of the elevator ramp marked the end of the Compact Strip Production machine: 1,177 feet from the mouth of the Kolakowski mold to the end of the ramp.

Iverson and Busse ended up among a sea of finished coils of steel that were sitting on the floor at the end of the ramp. The coils were 5 feet in diameter, weighing up to 23 tons apiece, and blue-black in color. Raw hot band steel, made from the Rust Belt. A great shimmer of heat was coming off the coils.

Iverson waved his hand over a coil, to see if it was hot, and then touched it, flicking his fingers lightly over the metal. He removed his safety glasses, hung them in his mouth, and tippy-tapped his fingers along the edges of the coil. "You are looking for that smooth edge," he said to me, his voice muffled by the eyeglasses. "Feel this one. See how smooth those edges are?"

The steel was blistering hot.

■

17

WAR ZONE

■ ▰ ■

A few minutes before seven o'clock on the evening of Wednesday, January 24, 1990, the casting crew brought a ladle over to the casting tower, filled a bathtub with molten steel, and prepared to cast, but they couldn't get the cast started. Mark Millett happened to be inside the pulpit of the casting tower at the time, along with Dave Smith, the casting foreman on duty. Smith, the helmsman, gave an order to terminate the cast. The crane operator, Mike Hinz, who was sitting inside a cab in the crane near the roof of the melt shop, moved the crane toward the casting tower to pick up the ladle and remove it from the casting tower—standard procedure when a cast is aborted. The ladle was nearly full. It contained ninety tons of hot metal, at a temperature of 2,900 degrees Fahrenheit.

On the casting deck there were three steelworkers: Jim Cotton, Greg Ward, and Terry Vanscoyoc. Pat Owens, the commissioning engineer, happened to be standing on the deck, watching the operation. Another steelworker, Dick Haltom, was inside the pulpit with Millett and Smith.

At 7:05 P.M., Hinz, the crane operator, lifted the ladle off the casting tower and carried it high and westward, toward a slagging area at the base of the tower—standard procedure.

Five seconds after the ladle cleared the casting tower, there was a whining sound from the crane. Millett, standing near the control desk inside the pulpit, heard the sound and looked up. He saw that the crane cables had broken and unraveled. The ladle was falling to the ground. It was a huge object, fifteen feet high, filled nearly to the brim with liquid

■

steel, and the bottom of it was forty feet off the floor. It seemed to pass the deck slowly as it fell, the crane cables singing in the winch. Five people were working on the floor below the ladle. Millett screamed with horror, thinking of the people on the floor. The ladle vanished below the deck. Millett and Smith both dove under the control desk. There was a big, bright, yellow flash, and the lights went out.

■ ■ ■

Three hundred yards from the melt shop, Franz Küper was working late in an office, catching up on paperwork. He heard a heavy boom. That was the ladle hitting the ground. A moment later a violent concussion rocked his office. At first Küper thought it was an airplane crash. His next thought was: *The casting tower is gone*.

He ran outdoors.

It was a black, moonless night, under an overcast sky. There were no lights in the melt shop, and the building was a black shadow as tall as Reims Cathedral. Flames were bursting through the building's walls, at the site of the casting tower. The building had exploded. He ran toward the melt shop.

■ ■ ■

Two hundred yards from the melt shop, a maintenance man in an office at the hot mill heard a boom and then a second boom, much louder than the first. He knew something had gone wrong at the casting tower. He ran outdoors. He looked southward toward the melt shop, along the line of the CSP, and saw a terrifying sight. He was looking through the melt shop. The building was a transparent spiderweb. The north and south walls were blown out, and parts of the roof were blown away. The casting tower was a silhouette bathed in orange fires at the center of the spiderweb. The blowouts in the walls were 60 feet tall and more than 120 feet wide. The main columns were intact, but most of the smaller structural steel was gone, and what was left formed a twisted mess in the flames. He thought something like: *Everyone in that building is dead*.

■ ■ ■

Keith Busse was at home when the telephone rang. A security guard was on the line. There had been a severe blast in the melt shop, possibly as many as forty people dead or injured. Busse dialed Iverson's house and got Martha Iverson, Ken's wife, on the telephone. She told him that Ken wasn't home yet. Busse told her that there had been a major accident at Crawfordsville. Then he called the Cotswold Building and learned that Iverson was on his way home in his car, and Iverson's car doesn't have a telephone. Keith Busse ran outdoors, got in his car, and headed for Crawfordsville at close to a hundred miles an hour.

■ ■ ■

The ladle hit the ground in a slagging area—a place where slag is dumped from ladles—and the molten steel inside the ladle lurched at impact, throwing steel everywhere in a chrysanthemum burst. The slagging ground was wet. It exploded. That blew out the south wall of the melt shop. The ladle tipped over against a retaining wall and dumped part of its contents on the ground at the base of the casting tower. The steel raced across the ground and fell into a tunnel beneath the casting tower. The tunnel had an open roof—it was more like a trench than a tunnel—and it may have contained water mixed with pieces of scrap steel. In any case, water was somewhere in the area. You can put water on steel, but you can't put steel on water. There followed an awesome explosion.

Inside the tunnel, molten steel buried the water and scrap, and the water flashed to superheated steam. Superheated steam, raised instantly to the temperature of molten steel, expands with a force that is a hundred thousand times greater than the pressure of the atmosphere. In other words, it produces a tremendous explosion. A fireball ballooned upward, throwing a fan-burst of liquid steel mixed with chunks of scrap all the way to the roof of the building, blowing out the north wall and parts of the roof. There followed a pause, because it takes a while for something to be hurled a hundred feet in the air and come down, and then molten steel and scrap metal came down over the casting tower, hitting the casting deck with great splats and clatters and back-splashes, landing on wooden pallets, bags of powder, wastebaskets, storage lockers, and cardboard boxes. The whole casting deck erupted in fires. Moments later, concrete in the deck began to explode under molten steel, throwing concrete shrapnel and liquid steel into the air again. Cans of cement exploded with thuds. Explosions missed all four men standing on the deck. The four men standing on the deck ran for cover and survived.

The spill and the explosion released a flash of infrared light. The flash filled the building with infrared radiation. The radiant heat inside the building rose to searing temperatures and then dropped quickly as the infrared light faded away. It was not unlike a nuclear explosion. During a

■

major steel spill, human skin can be blistered by blackbody light reflecting off walls and ceilings, as well as from the light coming directly from the metal.

■ ■ ■

Owens, the commissioning engineer standing on the casting deck, watched the ladle go down. He knew what to expect, and he thought he was going to die. In 1987, at the U.S. Steel Gary Works, on the Lake Shore, a full ladle of liquid iron was accidentally dumped on the floor of a large mill building. One man was drenched with iron and died. Owens visited the site of the Gary accident the day after it happened, and in his words, "It looked, Christ, it looked like a small nuclear weapon had gone off."

So as he watched the ladle fall, Owens knew what to expect: a tremendous release of heat, a great flash of light. He jumped into a corner below a girder and squeezed into the corner, trying to make his body as small a target as possible, feeling tiny, vulnerable, insignificant, a mouse in a house on fire. He felt a tremendous concussion. As the melt shop disintegrated and liquid steel monsooned across the deck but missed him, he felt as if something or someone were looking at him over his shoulder. *I got through one that time,* he thought.

■ ■ ■

Huddled beneath the control desk in the pulpit on the casting tower, Mark Millett and the casting foreman, Dave Smith, didn't really hear the explosion. To them it felt like a hurricane, and the metal rain came, and the deck caught fire. They crawled out from under the control desk and ran onto the deck into the fires. Smith was in charge of the casting tower. He was the pilot, with primary responsibility for the lives on the casting tower. Smith was twenty-nine years old, a tall, skinny, rather quiet person, with glasses, a narrow chin, and a loose-limbed way about him. He had worked for North Star Steel before he came to Nucor. At North Star he had once seen an electric arc furnace blow up. It was a chain of explosions that lasted for twenty minutes, while everyone stood around outdoors watching the melt shop disintegrate. As he put it, "I've been around Cape Horn a lot."

Smith and Millett dodged through steel fires and concrete explosions. Bits of concrete were flying through the air along with glowing pellets of steel. They wanted to get to the ground, because there had been people working on the floor of the melt shop. Smith turned around and said to his crew, who had moved out from under cover, "Call 911," and then Smith raced down the west exit stairs, with Millett running after him. At the base of the tower, Smith came to a doorway that led out onto the floor of the melt shop.

A brilliant light was streaming through the doorway. The light hurt Smith's eyes. He couldn't see anything through the doorway. It was like looking into the door of a furnace. He burst into tears, crying, "Charlie! Charlie!" He was thinking about a friend named Charlie Roarks, whom he knew had been working on the floor. He lowered his dark glasses. Now he could see. He went through the doorway into the light.

Through his dark glasses Smith saw the ladle in front of him, tipped at an angle, golden-mouthed, dribbling. Steel fires were burning everywhere. A natural-gas fire raged from a burst pipe. The space was filled with toxic smoke. Smith was wearing his silvers, but the radiant heat was so terrible that it seemed to cut through the graphite and Kevlar fibers in his suit. He put his hand over his face and tried to breathe, running through the fires, almost blinded with smoke and tears, crying, "Charlie!" He circled behind the ladle and looked there, "scared I'd find him curled up in a ball, just burned to death." His friend Charlie wasn't there. Nobody was there. One man had been blown fifty feet down the length of the melt shop.

Mark Millett followed at Smith's heels. Millett was not wearing silvers, and so the heat was for him all the more intense, if not life-threatening. He was a metallurgist, but he had never seen anything like this in his life. "The whole center aisle of the melt shop was covered in flames and molten steel," he later recalled. "It was a war zone." Millett covered his mouth with his hands and zigzagged around metal blobs, looking for survivors. There was no one in the area.

Charlie Roarks—the man whom Smith was hunting for—had been standing on the ladle preheat station, the small platform at the foot of the casting tower where ladles are heated and slidegates are repaired. He was not wearing silvers. When the ladle hit the ground, he tumbled down a flight of stairs, was caught in the steam explosion and the infrared flash, but he managed to get out of the building with only his hands and arms burned. Also standing on the platform was Patrick Major, the specialist from the North American Refractories Company who took care of slidegates. Major also was not wearing silvers, and he was overwhelmed in the heat.

A maintenance man named Eugene Ruark (no relation to Charlie Roarks) had been standing closest to the ladle when it fell. Gene Ruark was standing with his back to the ladle, talking on a telephone near the spot where the ladle hit the ground. The ground shuddered, and then he was pelleted across the back and shoulders by molten steel, which threw him to his knees. He was a Christian and accepted the idea that God might have chosen this moment for him to die. But he didn't die. He got to his feet and turned and took three steps. Then came the second blast, when the steel poured down into the tunnel and buried the water, and it picked up Ruark and hurled him fifty feet down the length of the melt shop. He landed conscious. He got up and started running, and as he ran he looked overhead, where he knew the danger was coming from.

What Ruark saw over his head he would remember for the rest of his life. He saw yellow blobs of liquid steel of different sizes, some like golf balls, many as large as water balloons, falling from above, dropping from as high as the roof of the building, a downpour of steel coming from perhaps a hundred feet in the air. The water balloons came down in slow motion. It took the metal four or five seconds to arc to the roof of the mill and come down. He knew that if he were hit by one water balloon it would kill him for sure. The golf balls missed him and the balloons splashed around him and he dodged them all and made it to safety behind a piece of equipment. His shirt was on fire, so he took it off. The steel had burned his shoulders but he had survived.

There were two other people on the ground, standing near a doorway that led outdoors. I will call them Philip Wood and Ellen Otway. When Philip Wood saw the ladle begin to fall, he said to Ellen Otway, "Get out." They ran outdoors, a wall blew apart behind them, and then Ellen Otway turned and ran back inside the melt shop to try to help, and she ran into the golf balls and water balloons.

Nothing hit Ellen Otway. The metal rain stopped, and she saw a mostly naked man, wearing only boots and gloves and scraps of burning clothes, running toward her, asking for help. It was Patrick Major, the North American man.

Major had been splashed directly with molten steel. He had taken the infrared flash and had been enveloped in a cloud of ash-dry superheated steam. The steam was hotter than the ignition point of cloth. The steam and the light burned his clothes off and left him naked, wearing only his boots and gloves. His hard hat was gone. Eighty percent of his body was burned.

The woman led him out of the building. His skin was red, shiny, and black, and marked with powdery carbon crusts from splashes of hot metal. He felt little or no pain, as yet. He was deeply afraid. Once outdoors they

■

stood in the darkness for a moment, and the woman took off her jacket and placed it over his shoulders.

The chief electrical engineer of the Crawfordsville Project, one Ralph Smith, came running to the scene. There wasn't much that he could do, but he tried to calm Major and removed a few bits of his charred clothing, and persuaded him to sit on the tailgate of a pickup truck. According to one report, when they tried to move Major, his skin slipped off like the skin of a persimmon. According to another report, he looked like he was wrapped in plastic.

The electrical engineer tried to comfort Major. He said, "Don't worry, Pat, when you get out of the hospital, I'll buy you a couple of beers."

Major began to laugh. "It hurts too much," he said. He complained that his hard hat was bothering him, could someone please remove his hard hat. He wasn't wearing a hard hat. His head was a hard hat.

Someone brought wet paper towels. They stuck the paper towels to his skin, to try to cool his skin. It was an abnormally warm night for January. Someone brought clean cloth towels, and they wrapped him in cloth and laid him down in the bed of the pickup truck.

As the steel fires began to die down, the melt shop lapsed into a dark cavern filled with smoke. No one knew who else might be inside the building. Steelworkers drove cars and pickup trucks up to the south wall and shone their headlights through the hole into the melt shop. The headlights penetrated the smoke, and then people began to walk into the melt shop.

■　■　■

Franz Küper entered the east end of the melt shop. Küper arrived at the foot of the casting tower. He saw fires burning on the top of the casting tower. At the base of the tower, he saw the broken gas pipe, and he saw how a gas fire lashed from the pipe. Already a knot of people had gathered near the casting tower. "Don't go up there!" they said to Küper. He borrowed a flashlight from someone and climbed the tower, where he found people milling around in darkness and flames on the casting deck, but everyone seemed to be reasonably calm on the deck, and there was no one to rescue.

Mark Millett and Dave Smith, the casting foreman, on the ground to the west of the tower, had by this time determined that there was no one near the ladle. Smith encountered a furnace foreman who told him that Charlie Roarks and Gene Ruark had escaped but that Pat Major had been severely burned. Approximately three minutes had passed since the explosion. Millett by this time had circled back to the base of the casting tower, where the gas fire raged. Smith ran back to Millett.

Millett said, "What about the crane operator?"

A sickening feeling came over Smith. He had forgotten about Mike Hinz, the crane operator. Smith looked up, overhead. The crane was motionless and wreathed in smoke.

Millett was carrying a flashlight. Smith borrowed the flashlight from Millett and gave Millett instructions as to how to shut off the gas fire. "Turn that valve," Smith said, and then he raced upstairs to the casting deck.

Mark Millett turned off the gas valve and many other valves. There were also oxygen pipes in the casting tower, and Millett feared an oxygen leak—if a cloud of pure oxygen erupted around the tower, it could trigger another fireball. Millett found a maintenance man and told him to turn off all utility lines running into the melt shop.

On the casting deck, Smith told one of his men to gather the crew together and make sure that everyone was accounted for. Smith felt guilty and humiliated. As he put it, "I had hired some of these guys. I basically took them out of a cornfield and put them into the steel industry." Smith headed for a ladder on the wall of the melt shop, to see what he could do about the crane operator, and he climbed the ladder toward the crane.

The ladder extended eighty-five feet off the floor of the melt shop. When he got to the top of the ladder, he was high off the ground and facing a crane rail. The traveling cranes are long steel bars that span the columns of the melt shop, and they move along crane rails, which are normally electrified. A glass-enclosed cab hangs from the bottom of the crane like a light bulb hanging from a girder. Since the power was down, the cranes were frozen and couldn't be moved. Hinz had been in the cab; the question was how to get to Hinz.

Smith got off the ladder and stood on the crane rail. The rail was dead; the power was down. He looked at the crane, which was a hundred feet from him, now. One of the cab's windows was shattered. Broken steel cables dangled down. A length of cable had flipped up and was lying across the crane rail—that was what had shorted out the power in the melt shop.

Smith began to walk along the crane rail. The rail is eighteen inches wide, and it looks straight down, eight stories, to the floor on either side.

■

It was from a crane rail like this that a Red River iron hanger had fallen to his death, bleeding from his eyes after he had hit the ground.

From where Smith stood on the crane rail, he was looking down into the open tunnel that had just blown up. The tunnel contained puddled molten steel, and a second explosion from the tunnel, or from the slagging area, was by no means out of the question. The building was smoky and without illumination except for the light of glowing steel and various wildfires. The gas fire withered down—Millett had managed to shut off the gas. Smith continued to walk along the crane rail. He had to walk a hundred feet to reach the crane. The ladle, leaning against the retaining wall, was partly full of steel, and was plastered with slag and skulls. Fires were sizzling in patches over the floor and walls, and paint, plastic, wires, and bags of material were burning, sending up columns of toxic smoke that pooled near the roof. It was dark and poisonous up there.

"The breathing was getting pretty bad," in Smith's words. "There was all this stuff burning, and my throat was hurting bad. The blast had blown a window out of the crane. So I started getting kind of scared. I didn't see any movement in the cab."

He reached the crane, climbed onto a catwalk, and moved along the top of the crane. "I thought, 'I'm going to climb down in there and find someone in there hurt real bad.' I got my flashlight out and shined it in there. Then I heard him hollering."

Hinz, the crane operator, was the man who had once mentioned to me his dream, that he was waiting for his retirement plan to hit six figures so that he could buy a mobile home and retire. The concussion broke the window of the crane cab and threw him to the cab's floor, covering him with broken glass, but fortunately the cab did not take a direct hit of molten steel. Hinz had climbed out of the cab and was now sitting on a platform on the side of the crane, with an injured shoulder, suffering from shock and smoke inhalation.

"Mike was saying 'Dave,' and everything," in Smith's words.

Smith led Hinz along the crane and back to the crane rail. Now they had to get back to the ladder on the wall, along the hundred feet of rail, and Hinz was dizzy. Smith held Hinz tightly. He told Hinz to sit down on the rail if he felt faint. They began to move along the crane rail toward the ladder. If either man had lost his balance they both would have pitched to their deaths. They reached the ladder, and Smith got Hinz down the ladder to the casting deck. Mark Millett, meanwhile, gave orders for all the

■

employees in the melt shop to gather by the melt shop for a head count. Millett did two head counts. Everyone seemed accounted for.

■ ■ ■

People in Crawfordsville listen to police scanners. Early accident reports indicated that as many as forty people might be dead or injured at the Nucor plant, and that caused a general panic in Montgomery County. People from all over the county got into their cars and headed for the plant. Soon the roads leading to the Project were streams of onrushing headlights. Many employees had parents living in the area, and the parents naturally headed for the plant. Within fifteen minutes, two hundred cars had gathered in a traffic jam outside the north gate of the plant. People could see in the distance the black, broken, faintly glowing melt shop. In the traffic jam some people were weeping, a few were begging to be let through the gates, but many people seemed to be in shock, while all around the melt shop flashed lights of ambulances and fire engines.

Iverson stayed on the telephone, alternating between Keith Busse (Busse was in his car, heading for the mill) and various people at the mill. During the first minutes after the accident, nobody knew the number of casualties, and it looked to Iverson as though this might turn out to be the worst accident in the history of the company, worse than the 1974 boilover on the Darlington casting deck when four Nucor steelworkers died. Iverson's conjecture was wrong; in fact, no one had yet died.

Janice Roach, the safety director, had just arrived home from work when she received a telephone call from a security guard at the plant. He said that the melt shop had blown up and that there might be bodies all over the place. She told the guard to call her safety assistant, and tell the assistant to go straight to Culver Union Hospital in Crawfordsville, because Nucor employees were going to be arriving there in ambulances. Then she drove to the plant, and when she arrived, she had to work her way through the traffic jam. She found that the injured had been taken to Culver Union Hospital. Roarks and Ruark were being treated for burns, and a third man, Dick Haltom, who had been on the casting tower, was being treated for smoke inhalation, as was Mike Hinz, the crane operator. Pat Majors had been fitted with intravenous drips to combat shock and had been evacuated by Life Line helicopter to Methodist Hospital in Indianapolis, where he was put into a sterile environment in a burn unit. Majors suffered from first-, second-, and third-degree burns—red burns, watery blisters,

■

and charring. She didn't think he was going to live. In her words, "You can't have eighty percent of your body burnt and have much chance of survival."

■ ■ ■

Manfred Kolakowski was in his room at the General Lew Wallace motel when the telephone rang, and somebody told him that there had been a terrible explosion in the melt shop. Kolakowski's first thought was that the casting machine had detonated. He thought that molten steel had somehow mixed with water inside the machine, triggering a fireball. Kolakowski now insists that he never doubted that the machine itself had any fundamental flaw in its design, but he believed, then, that something unlucky had occurred inside the machine. He arrived minutes after the explosion at the plant, where he found the crowd of cars already gathering by the gate, and then the ambulances started to come through. But the security guards wouldn't allow him inside the plant. "I'm the inventor of the machine," he told them, but they still wouldn't let him in. So he drove back to his motel room and waited for news.

■ ■ ■

Keith Busse drove through the south gate, where a guard told him, "Mr. Iverson's trying to get hold of you." Busse took a quick look around and called Iverson again on the telephone. According to Iverson, Busse sounded "scared as hell." As Iverson put it, "Keith had never been through anything like that, and it was a terrible thing for a manager to go through, a test of a manager. At a time like that, you don't know which way to turn, and you are frightened. Your employees are frightened themselves, and what do you tell them?"

In Busse's words, "Until you see the walls of a steel mill blown off and part of the roof blown away, the power of hot metal doesn't hit you. You come face to face with hot metal." Busse spent most of the night at the plant, walking inside and around the melt shop, talking with steelworkers, trying to calm them down, trying to find out what happened. He checked into the Holiday Inn at three o'clock in the morning, slept for a couple of hours, and returned to the plant before dawn to try to begin to put the Crawfordsville Project back together.

■ ■ ■

■

Janice Roach spent the night dealing with fire officials and medical personnel, and answering questions from newspaper reporters. The firemen turned on their hoses and sent streams of water in through the melt shop's walls all night. The water boiled when it hit the floor, and clouds of steam came out of the melt shop. By sunrise all of the ground inside the melt shop was cool enough to walk on. Around four o'clock in the morning, after the reporters had gone, Jan Roach found a quiet place for herself and began to cry. She cried for a while and then she stopped crying. After that she didn't talk about the accident with her children, friends, or relatives. She was divorced and was seeing a gentleman, but she didn't talk about it with her friend, either. "I just wanted to carry it for a while."

■ ■ ■

In the Scoreboard Lounge, near Main Street in Crawfordsville, under a glowing sign that looked like a pillow, inviting seekers of liquor and a game of pool to enter and make themselves at home, Patrick Owens, the commissioning engineer, sat alone at the bar. He drank eighteen shots of straight tequila over a period of several hours, enough tequila to fill a bottle and then some. That's what he did for a living, he commissioned casting machines. He knew what molten steel could do, because he'd seen it on the Lake Shore. And he and a lot of other people at Nucor were very, very lucky that night, because they were alive. He closed the bar down.

■ ■ ■

In Charlotte, North Carolina, in his house beside a lake, F. Kenneth Iverson could not fall asleep. He slipped out of bed in the darkness, careful not to wake up Martha. He padded in his pajamas to a nearby room. He flipped on a light and sat in a chair beside a window overlooking the lake. The lake was gray and flat under floodlights that burned all night. He sat up much of the night in the chair by the window, not even trying to get to sleep. He looked across the lake at the wood ducks, mandarin ducks, Canada geese, and black swans. The water birds floated on the lake, asleep in the water, their backs silvered with frost, drifting with their heads under their wings, or sometimes they moved restlessly. If a raccoon crossed his yard, he wondered what was going through the raccoon's mind—probably trying to eat one of his ducks. To watch the black swans from the window while he thought about the company helped him to calm down. The Nucor Corporation was Iverson's life. Therefore in some respects the company

■

was Ken Iverson. In other respects the Nucor Corporation was no longer Ken Iverson, because a corporation has a life of its own. In the early hours of the morning he padded back to bed and fell into confused dreams. He had lots of dreams. Whatever they were, they weren't about hot metal, because he said he never dreamed about hot metal. His dreams took other forms, not exactly about steel mills, but somehow they were all about the company and the people in the company, all mixed up into crazy situations, although he couldn't remember them afterward. Even when he slept the Corporation was his life.

■ ■ ■

If the ladle had fallen about five seconds earlier, it would have landed on the casting deck and drenched the casting tower with molten steel. It would have burned up the casting tower, and in that case it is generally believed that no one on the casting tower would have survived, because no one would have had time to escape. Numbered among the dead would have been Mark Millett.

Millett had come within five seconds of losing his life. Among those who understood how close a call it had been for Millett and the other men on the casting tower was Abby Millett, Mark's wife. "Abby was petrified," said Mark. Other spouses did not want their husbands or wives to return to work at the steel mill.

Inspectors from the Occupational Safety and Health Administration, OSHA, came through the plant and sifted through twisted metal and broken crane cables. Some of the crane cables had fallen into the ladle of molten steel and had dissolved, and so crucial evidence was gone.

It might have been possible to restart steel production within a day or two—although the walls were blown out, the machinery was for the most part undamaged—but Busse didn't want to make steel. The OSHA people found various safety violations. They reported, "The employer did not furnish a place of employment which was free from recognized hazards that were causing or likely to cause death." OSHA levied a thirty thousand dollar fine on Nucor; Nucor is contesting it.

Busse shut down production and organized a series of meetings with employees. From Busse's point of view, the purpose of the meetings was to allow employees to express their complaints and to offer ideas for ways to improve safety. "I pounded the pavement," as he expressed it. "You start with a series of safety meetings. You don't have one safety meeting, that would be a brush-off. We also tried to demonstrate the safety of the

■

mill to them and to present our employees with facts and figures that show they are working for a safe company." There was a change in Busse. He had tasted the bitter apple, and the heady feeling of playing with gigantic toys had given way to a deepening awareness of the possibility of death inside steel mills. Busse had come to earth in a severe bump.

The employees from the local area were horrified by the accident. They quietly expressed a great deal of fear during the meetings with Busse, and they pointedly asked him if the Corporation cared for them. Crawfordsville had woken up to steel. Nucor employees who had come from other companies in the steel industry felt that a warning had been given to the inexperienced Nucor employees, and they also expressed the sentiment that Busse behaved himself well; that there was no cover-up and no witch hunt; that Busse listened, and that Busse appeared to be as shocked by the accident as the local people. A former employee of U.S. Steel put it this way: "The accident woke up a lot of people to the magnitude of the dangers in a steel mill. It woke up Keith Busse. He told us that he himself had walked underneath ladles without really thinking of what was over his head. But I've been trained, and when I see a ladle coming, I go around a corner and peer at it."

Busse met with employees in small groups, and then in larger groups, and finally in a series of general meetings. After a week of meetings, Busse restarted production. He vowed to make the melt shop the safest melt shop in the United States, and he announced a series of changes: double cables on the cranes, so that if one cable breaks, the other cable will prevent a ladle from falling to the floor; collapsible escape ladders that can be thrown over the railing of the casting deck, to get people off the ship if it's burning; and a rule that all employees who are working in the aisle of the melt shop have to go outdoors when a ladle is coming through. He said he would build a catwalk running the length of the melt shop, so that employees could walk the length of the melt shop without ever having to set foot on the ground.

Dave Smith, the casting foreman who had climbed to the roof of the melt shop to save the crane operator, had this to say: "People in the steel industry kind of take these accidents as a fact of life, and Keith didn't because he hasn't been around steel. Keith's talking about spending a lot of money on safety that the law doesn't require him to spend, like putting double cables on the cranes. If you wanted to be completely safe, you wouldn't drive a car. Cars are dangerous, and most people drive cars. Some people think cars are too dangerous, and won't drive them. Other

■

people ride motorcycles without a helmet. The only way to prevent all accidents with steel is never to have a ladle full of steel.''

The specific cause of the accident remained a mystery to OSHA. The cable may have snapped. It may have broken because of abuse—the crane operators may have been working the cranes too hard and too fast under the pressure of the startup. Keith Busse denies that possibility. He says that the crane operators were following safe procedures, but a strand of crane cable pulled out of a clamp, allowing the ladle to fall to the floor. The disaster may well have happened because of one loose clamp.

The clamp, known as a "dead-end" clamp, gripped the cable. The clamp was held in place by four bolts. The cable was one inch in diameter. It may have stretched in use, shrinking by about an eighth of an inch, until it suddenly slipped out of the clamp, letting the ladle fall to the floor. A clamp held by four bolts had loosened up, and the east end of the melt shop had exploded. But industrial accidents never happen for simple reasons. They happen when a linked series of errors builds toward an event, which in this case was a ladle hitting the floor.

Driven by the pressure of the startup, Busse and the Nucor employees evidently did not pay enough attention to routine maintenance. In particular, OSHA requires that all ladle cranes be inspected weekly for safety, and OSHA also requires the company to keep records of the inspections. Nucor did not inspect the cranes regularly and did not keep records. That was a serious violation of safety codes. In Busse's words, "We had rules for maintenance, but in the thick of battle we didn't enforce the rules. If we didn't get to something, we'd say, 'We'll get to it next week.' " The Nucor maintenance crews were busy and inexperienced, and they didn't monitor the crane cable clamps. Busse, realizing that his maintenance crews were overworked, had hired a licensed crane inspection firm to inspect the cranes for safety. In Busse's words, "We hired an outfit to come in and inspect our cranes. They had just inspected that crane! They had inspected it on December 26 [one month before the accident]. Possibly we will find mechanical or design failures in that crane, I don't know. The crane may have been rigged wrong to begin with—we think the dead ends [the cable clamps] may have been improperly torqued.'' In other words, the cable clamp bolts may not have been tightened properly after the crane was installed in the melt shop, according to Busse's claim.

The ladle fell into a slagging area—the best possible place for a ladle to fall. But there happened to be water sitting somewhere near the casting

■

tower. OSHA charged Nucor with a knowing violation of safety for having allowed water to accumulate near the casting tower and in the slagging area. Nucor is contesting that charge. What seems most reasonable is that the explosion was caused by water flashing to superheated steam underneath liquid steel, throwing the steel into the air at a hundred thousand atmospheres of pressure in a devastating blast. The steel found water and went through the roof, as Mark Millet had once predicted could happen if you put steel on water.

Patrick Major, burned over eighty percent of his body, went in and out of his mind in the hospital in Indianapolis, pumped full of morphine, and yet still it hurt too much. It is said that to die slowly of burns is among the most painful of deaths. Burn victims have terrible problems with electrolyte balance, because their skin oozes and keeps losing moisture. But even though their skin is wet they can't sweat, and so they can't control their body temperature, and they suffer from raging fevers. Essentially the skin is one large blister. Spreading bacterial infections rotted Major, and his kidneys slowly failed.

Mark Millett and a couple of other Nucor people went to the hospital to see Major. In a waiting room near the burn unit, Millett encountered Major's wife, grandmother, sister, and a lawyer representing the Major family. Millett said to them, "We're from Nucor, we came to see how Pat was doing." Millett found the Major family to be bearing itself with dignity. He also found the family to be extremely uncomfortable in his presence.

The lawyer asked Millett, "How did it happen?"

Millett tried to explain how it had happened. Then he went into the burn unit, where he saw what was left of Pat Major wrapped in sterile white sheets, and came out of the room shaken.

Major died a few days later, at the age of thirty-seven. As one Nucor employee put it, "Pat would have been better off if he had died inside the melt shop." Shortly afterward, the lawyer representing the Major family indicated that his clients would seek large, unspecified damages from the Nucor Corporation. The matter is now in negotiation among a number of insurance companies. One way or another, the Major family will receive a sum of money in exchange for the life of a man. In the Middle Ages it was called *wergeld*: man-gold. He had been a pleasant man with sandy brown, thinning hair, an expert on sapphire slidegates, who worked hard and enjoyed a beer at the Scoreboard Lounge. He was survived by his wife and two children. Millett attended the fu-

■

neral, in Valparaiso, Indiana, near the Lake Shore. The children wept inconsolably and Millett drew many stares, some bitter, some sympathetic, some curious. Imperceptibly, Mark Millett had become the Corporation.

18

THE RIVER

■ ■ ■

O n a fine cold day in April 1990, three months after the accident, at the
Cotswold Building, on the outskirts of the city of Charlotte, North
Carolina, F. Kenneth Iverson was sitting at his desk in a pool of
banded light. His venetian blinds were open, for a change.

"If you've never been in a steel accident, it's really a traumatic experience," he said, and his gray eyes widened. "Keith was magnificent. He
handled it beautifully. Nobody so far has left the company, but there were
people who were close to leaving. Keith took a lot of time with people and
allowed everybody to have their voice."

Ellen Otway, the woman who had led Patrick Majors naked in his boots
out of the melt shop, had been on the verge of quitting. It seemed evident
that the chairman had been quite well aware of her trauma and had been
fearful that she would indeed quit, which would have been a stinging blow
to the Corporation if not to Iverson, in some symbolic way.

He stood up and crossed his office and picked up a metal ring off a shelf.
He dropped the ring on his desk with a clatter. "Take a look at that," he
said.

I held the ring in my hands. It was a piece of silvery gray steel, marked
with the corporate logo of the Ford Motor Company.

"That's a bracket for a car stereo speaker. That's hot band steel, from
Crawfordsville," he said, sitting down. "We're selling steel to a parts-
maker who's selling these to Ford. I don't think the Ford Motor Company
even knows that it's putting Nucor steel into Ford cars." He grinned. He

■

crushed a cigarette in a dusty soapstone ashtray, stood up, and reached behind his door for his jacket. We passed the starveling umbrella tree and the receptionist, and rode down in the elevator. Iverson headed across the parking lot and slid behind the wheel of a custard-yellow Mercedes Benz turbo-diesel. The car was in less-than-mint condition, with a greasy leather interior, and it smelled faintly of tobacco. He pulled into traffic beside the shopping center, keeping to the right lane and driving fast.

"There is always a cost of market entry," he said. "We're selling at twenty dollars a ton lower than Big Steel's price, which is currently about three hundred and twenty-five dollars a ton. Big Steel is making it difficult for us to make a market entry."

In other words, Keith Busse was locked in another gun battle for market share, this time with the big boys on the Lake Shore.

He floored his turbo Mercedes. You could hear the blowers kick in, and he passed a car on the right.

"Ha! I remember one time"—raising his voice—"George Cannon, Senior, at Cannon Muskegon, wanted to break into the market for cast-iron furnace grates. Iron furnace grates were then selling for twenty cents a pound. So Cannon offered iron grates for *eighteen* cents a pound. But then he had to figure out how to *make* an iron grate that he could sell for eighteen cents a pound and still give him a profit. What he did was, he mixed a lot of slag in with the iron. He got a hell of a hard grate, it was a beautiful grate, except you didn't want to drop it or it would shatter like a piece of glass. I can still feel him grabbing me by the back of the arm—'Ken, a good businessman is hard to bruise and quick to heal.' "

He swung down a narrow lane, through woods shining with azaleas, past houses settled in loblolly pines. The road wound down a hill. He pulled into his driveway, into an uproar of geese honking behind the house—*hoooak! hoooo-ak!* The house was low, wide, made of brick, with pyramidal roofs. We entered through the front door. "Hello, Caspar!" he said to an African parrot perched in a cage in the front hall. Caspar didn't reply. Caspar was sullen today. Caspar scraped his beak on his perch and glared at Iverson.

Martha Iverson was a handsome lady, with silver earrings, bobbed hair, and a forthright look on her face. "It is *so* cold today, and this is supposed to be April," she said. "I am not one of those people who thrives in the cold."

Ken went into the kitchen and pulled a head of iceberg lettuce out of the refrigerator, and in a cabinet he found a box of Saltines. He went through a glass door into the back yard, and stopped at a gate in a wire fence. He broke up the lettuce and threw it in chunks over the gate. *"Hulloo! Hullooooo!"* he cried.

The birds heard him. They answered him with excited cries.

He went through the gate and down to the lake, and stood on a dock. "Jessy!" he shouted.

The black swans were sitting on the far side of the water, on the banks of a greeny-brown pond. The birds hurried into the water and started paddling toward Iverson.

"Jessy! Where are you, girl? Jessica! Jessy, baby! Hello, Jessy! Pretty girl! *Uh-hoooo!* Black swans are gregarious. That's why I like 'em. They're the only swans that are friendly. *Huh-ooo!* Come on!"

Jessica was a large black swan, with a red bill marked with a white ring around it. He shredded Saltines and flung them in the water. Carp rose on the Saltines. Three small ducks hurried away from the shore, fleeing Iverson's voice. "Those are mandarin ducks, the most beautiful duck in the world," he said. "They're very shy." Three dignified homely water birds paddled around the dock. The birds had bulbs on their beaks. "These are what they call Chinese geese. Where's Mrs. Redlegs? *Uheeeoh!* Where *is* Mrs. Redlegs?"

The geese answered Iverson with quiet voices, nibbling on Saltines. A line of dogwoods on the far side of the lake was coming into bloom.

We ate lunch with Martha Iverson—"I'm not worried about Ken around steel," she remarked, over backfin crab, "I'm worried about the way Ken drives a car." Ken had to return to work, and as he was going out the door, Caspar spoke. The bird said, "Bye, bye! Where are you, Ken?"

He drove fast back to the office, hugging the right side of the road, threatening to clip mailboxes. He parked beside the Cotswold Building and headed for the front door.

I asked him if people at the company were worried about what might happen to Nucor after he was gone.

"I guess the worry is that the company will buy an airplane and put in a lot of group vice-presidents." He shrugged. "There'd be a reaction from the general managers if any CEO of this company tried that. But I'm replaceable. Any number of people could take my place and maybe do a better job." He paused to admire the Cotswold Building. The structure's buff-colored bricks glowed in afternoon light, bright with aluminum detail.

■

"We're up to twenty people now," he said. "We got 'em all fitted in here!"

■ ■ ■

Companies are clouds, moving together in the general flow of the national economy. You think you see a shape, but the next time you look the cloud is different. The last time I looked at the Nucor Corporation, it was becoming a thunderhead. The Crawfordsville Project was a financial success. The mill had turned a profit, and customer orders were booked for several months in advance.

"We turned a profit in ten months, and that's unheard-of for the startup of a steel mill," said Keith Earl Busse. "Sleeping Beauty woke up after all."

The mill was running at a tremendous rate of speed. The Compact Strip Production machine was cranking out steel at a rate of 900,000 tons a year, operating at 90 percent of its projected one-million-ton capacity, but now it looked like the mill's capacity would probably exceed 1 million tons a year. Busse believed that he might be able to goose Iverson's machine well beyond a million tons a year. He thought that he could increase the rate of fire to 1.3 million tons a year. Wall Street's engine runs on the emotion that comes with the gain of money or the loss thereof, and Nucor's stock was trading fast and loose on the New York Stock Exchange. One day it fell 6⅞ points, another day it leaped 6 points. Last time I looked, it was at $63 a share. Sam Siegel, Nucor's chief financial officer, estimated the capital cost of the Crawfordsville Project at $275 million, and he claimed that it was a bull's-eye on his original estimate, after he had factored in Iverson's usual budgetary optimism.

Keith Busse punched some numbers on a calculator and declared that Iverson's machine had most recently made hot band steel with 0.66 hour of human labor per ton, and that included all administrative staff. Two-thirds of a man-hour per ton. Iverson's machine had reached well beyond the labor efficiency that Iverson had originally projected, which was one man-hour a ton. The Crawfordsville steel mill was six times more labor-efficient than the Japanese steel industry.

Keith Busse was selling steel for as little as $280 a ton, 14 cents a pound. For virtually all steel companies on earth, fourteen-cent steel is the kiss of bankruptcy, the valley of the shadow of death. Busse was selling metal into the dark valley and he was earning a profit there, and he feared no recession. He almost looked forward to a recession, because he believed that the next time the American economy got sick, "Big Steel will be painting the walls red with their blood while we may be eking out a profit," as he put it. The name of the game is survival.

But there was a reason why Busse had to sell some of his steel for 14 cents a pound. It suffered from brown spots. Some of it was Distressed Material. He believed that within six months he could match the quality of the sheet steel coming out of Lake Shore, and could consequently get a better price for Nucor steel.

Ken Kinsey, the heavyset fellow with gentle eyes behind eyeglasses, whom the iron hanger had threatened to kill, was promoted to be a foreman on the casting tower. Meanwhile, the Nucor Corporation had crashed into the market for skyscrapers—the Arkansas steel mill, under the leadership of John D. Correnti, had seized twenty percent of the American market for I-beams, and the Arkansas mill was now selling steel to Indonesia and Japan. American steel was moving west on ships to the Pacific Rim. John D. Correnti and Keith Earl Busse, both with steel mills under the belts, were watching each other more closely than ever. Iverson had announced plans to build Nucor II, the dreaded clone of the Compact Strip Production machine. Nucor II would be built in Arkansas. He had negotiated a deal for the clone with SMS. But he had not yet named a general manager of Nucor II. It was rumored that John Correnti would be made the general manager of Nucor II and that Correnti would try to crash-build Nucor II even faster than Busse had built the Crawfordsville Project.

The Koreans, the Germans, the Belgians, the Mexicans, the Canadians, and the Brazilians, none of whom were collaborating with Nucor but all of whom were exporting steel to the United States, were scared, according to Iverson. In his words, "There is no question that the foreigners are afraid of us." In Iverson's opinion, the foreigners were afraid that if Nucor kept building CSPs, Nucor might hurt their export markets in America. The foreign steel producers were worried that they might lose a piece of the five Great Pyramids of solid steel that Americans consume every year.

■ ■ ■

While the Nucor Corporation made more and more steel, differing views on the subject of Nucor, formerly the Nuclear Corporation of America, formerly the Reo Motor Car Company, continued to be heard. "Nucor

■

could lose something," said one knowledgeable observer in Colchester, Vermont, at the headquarters of the Hazelett Strip-Casting Corporation, which had supplied an experimental casting machine to Nucor. "It reminds me of what happened to People Express. You remember People Express, the airline? They were headquartered right here in Vermont. They were like a family when they started out, and they got too big. They grew to the point where the employees no longer knew each other. And where are they now?"

Keith Busse had a different idea. "When we push the 'Run' button, it's hang onto your ass," he said.

Joseph Kinney, the director of the National Safe Workplace Institute, remarked, "I hope Nucor gets the justice it deserves."

David A. Thomas, the former chairman and president of the Nuclear Corporation of America, who resigned from the company in 1965, felt that he had not received the credit he deserved for the success of the Crawfordsville Project. He was very glad, however, to see that Iverson had stayed the course. As Thomas saw it, the Crawfordsville Project was the inevitable result of his own chairmanship of the Nuclear Corporation of America. "It would be exactly the same thing if I had stayed with the company," he said. "They've successfully followed the program that I had already launched. Of course I would have done the same thing in Indiana! *Of course* I would have! Because that's the way it was destined to go!"

A senior vice-president at Bethlehem Steel, Roger P. Penny, was quoted in the *New York Times* as saying, "To make their product fully competitive in the marketplace will surely take five to ten years."

"That's his opinion!" fumed Busse. "He's never seen Crawfordsville! He's never stumbled over this technology in his life! What we're doing in Crawfordsville is like taking a Conestoga wagon for the first time across the plains."

Not long ago F. Kenneth Iverson received a letter from Roger P. Penny, senior vice-president, Bethlehem Steel, Bethlehem, Pennsylvania, asking if Mr. Penny could visit the Crawfordsville Project to inspect Iverson's machine. Iverson was happy to oblige Mr. Penny.

Dick Haltom, a Nucor steelworker who had been on the casting tower when the melt shop blew up, was asked by a reporter if he was satisfied that the mill was a safe place to work. He answered that he was personally satisfied that the mill was safe. He had been around manufacturing plants before, and in his opinion the steel mill was safer than a wire mill in

Crawfordsville where he had worked before he came to Nucor. He added, "We're not making aspirin here. We're making steel."

■ ■ ■

David C. Thompson, hot metal man of the second generation, stands on the casting deck, suiting up for a cast. Big Dave wraps his silver booties around his legs. He fastens his silver coat. His silvers are filthy, greasy, eroded by the violence of the startup. Thompson carries a flashlight in his hip pocket for peering into lightless nooks inside the casting machine. A siren wails through the melt shop: a ladle of steel is heading for the tower. The ladle arrives at the deck, is placed on a ladle car, and it creaks on rails toward the copper mouth. He turns and gazes up at the ladle, his spectacles glint, and he lowers his faceshield in preparation for the duel.

A drumroll splits the air, an arc furnace melting at full power. Keith Earl Busse can often be found on the furnace deck. When he's there, he takes a look around to see that the area is healthy, safe, and clean, he says. ("Is this smoke being pulled away properly?") After that, inevitably, his eyes turn toward that little door in the side of the furnace that leads to another universe, and he lowers his dark glasses. "I guess there's something so God damned exciting about living on the edge. I guess I'm a little awe-struck by the power and the energy that's sitting there, contained in that flat bath. When I see those electrodes popping the metal, I think about what kind of awesome energy is being driven into a square meter of metal. Snap, crackle, pop, and bang. I love to hear arc furnaces roar. In the business world, you live where the excitement is. I think the steel business is the most exciting business I've ever seen in my life. I guess I'm a hot metal man at heart."

At the casting tower, Manfred Kolakowski paces the deck. "I have not been back to Germany since I arrived here," remarks Kolakowski.

Big Dave Thompson says, "Cast mode," and the sequencing begins. Hot metal men from Crawfordsville move quickly over the deck, setting up temperature probes, throwing levers, and dumping powder beside the mold.

"Open slidegate," says Thompson.

The ladle opens but nothing comes through. The slidegate is plugged with a skull. Thompson takes up an oxygen lance and drives it up into the ladle. He yanks out the lance, there's a flash, steel drops into the bathtub in a brilliant pour, and Thompson turns away from the hot metal stream without a word. His crew knows what to do. The bathtub fills with liquid

■

steel. They throw bags of rice hulls into the bathtub and flames erupt from the bathtub. Steel rushes into the Kolakowski mold, which shivers around the nozzle in a sexual-industrial climax, and water jets roar inside the machine.

A steelworker calls to Thompson, "We're casting at a hundred and seventy inches a minute."

Thompson studies a control panel. "I've got seventy tons of steel in the ladle," he calls out. "Slow it down a little, the steel's about to blow us out of here."

The slab emerges from the base of the casting tower, a lengthening piece of pasta. The slab rides slowly through the tunnel furnace, being heated to a golden glow, heading for the millstands. At the pulpit of the hot mill, a foreman watches symbols move across a colored screen. He mutters, "Yeah. Here it comes. Here it is. Ready?"

An operator nods, his fingers poised over joysticks.

The slab passes through a water spray with a great puff of steam and drives into the roll bites—*ba-boom, ba-boom, ba-boom, ba-boom*, four thunderclaps as the steel is squeezed through the mill train. It sounds like a lightning storm.

The strip of steel, now three thousand feet long and a tenth of an inch thick and still glowing, tails out of the mill, hurries through water jets, darkening to gunmetal blue as it cools, and as the tail of the strip rumbles into the downcoiler, a final *ba-boom* echoes through the mill.

In a pulpit opposite the downcoiler, at the butt end of the hot mill, at the most boring place in the entire Crawfordsville Project, Will W. Hawley, distinguished former Run-a-Mucker, unguided missile, ex-builder of condominiums, erstwhile dealer in guns, survivor of a crane drop, drinker of two consecutive Nucors without throwing up, part-owner of a farm in Ladoga, and builder of a steel mill, is smoking a five-dollar cigar and punching the occasional button. Hawley seems to be dismayed.

"I'm chomping at the bit," he says. "I'm feeling kind of left out. Eventually I will be replaced by a robot. I like the fast-hitting jobs, where you carry on three conversations at one time. It was much more exciting building this mill than it is actually running it." Hawley had worked at the job site practically since the bulldozers had hit the land, he had logged sixty-hour weeks to restore American steel to its former greatness, and now, somewhat to his surprise, he is a steelworker.

Another slab noses into the mill train, *ba-boom, ba-boom, ba-boom, ba-boom,* and a strip is squeezed from the mills and hurries down the line

■

and hits the downcoiler, *ba-boom*. Hawley's hard hat turns as he watches a coil of steel pop off a mandrel. The coil moves slowly up the ramp to the end of the line. Iverson's machine carries the river and the river shines like the sun, flowing from the mountains of the Rust Belt to the arc furnaces of Crawfordsville, down the cataracts of the casting tower, through the narrows of the rolling mill, and into the Midwest.

CREDITS

■

I owe first thanks to the employees of the Nucor Corporation. They welcomed my presence as a reporter among them, while they built and started up a steel mill. It is not possible to name them all, for the simple reason that if I tried to list everyone who helped me I would fill up several pages but would still manage to forget one or two names, and I wouldn't want to do that. I owe thanks to the many employees of SMS Schloemann-Siemag A.G., of Düsseldorf and Hilchenbach, Germany, who welcomed an American reporter among them.

There are people associated with Nucor who were kind to me in certain special ways, whose names in some cases do not appear in this book, and I wish to acknowledge their kindness here by name: Carol Busse, Jim Clinger, Bill Dauksch, Norma Furnish and Kenneth Neuhouser, Will W. Hawley, Jim and Dianne Hoskins, Martha Iverson, Marsha McDonald, Kris McGee, Abby Millett, Roy Shrout, and Ralph and Carolyn Smith.

Various employees of SMS Schloemann-Siemag A.G. gave particularly valuable technical help: Angelika Demmer, Franz Küper, Manfred Kolakowski, Hans-Friedrich Marten, Patrick Owens (with SMS Concast, Inc.), and Hans Streubel. Dieter Richstein of SMS gave technical help of a considerably more precise nature than the application of *Korn*.

For help with the birth of my daughter, which occurred during the early morning hours of June 17, 1989: Sheila Jasento, of Deluxe Travel Bureau, Inc., Lollie and Richard O'Brien, Dolores Parham, John and Diana Parham, Jerome Preston, Jr., Donald P. Schneider, Daniel W. Shapiro, M.D., Karen Shenk, R.N., Edwin and Joyce Turner, and Ann Waldron. They come when they're ready to be born.

I wish to thank Jerome Preston, Sr., for continuing financial help. Many thanks to Jet de Zeeuw Schneider for fixing my abysmal German. Alan P. Lightman gave valuable editorial hints and also helped fix certain statements involving physics. For literary advice and assistance: John Bennet,

■

Susan Danoff, Frederic D. Grant, Jr., Chip McGrath, John McPhee, David G. Preston, M.D., Douglas J. Preston, Michael Robertson, and Brian Abel Ragen. Thanks also to: Mrs. Richard H. McCann, Wilfried and Karin Eckstein, Mrs. Eugene Stanley, and Anna Marguerite McCann Taggart and Robert D. Taggart.

For a combination of friendship and keen editorial judgment, I am grateful to Paul Aron, Executive Editor, Prentice Hall Press. For their sincere expressions of faith in me, Elizabeth Perle, Publisher, and Marilyn Abraham, Editor-in-Chief, Prentice Hall Press. And many thanks to Sallie Gouverneur.

For everything, my wife, Michelle Parham Preston.